Basic Horticulture and Fruit Production Technology

NIPA® GENX ELECTRONIC RESOURCES & SOLUTIONS P. LTD.
New Delhi-110 034

About the Authors

Dr. Rajaneesh Singh is presently working as Associate Professor & Head, Department of Horticulture, Tilak Dhari Post Graduate College, Jaunpur (U.P.). He did his Graduation in Agriculture from Udai Pratap Autonomous College, Varanasi, Post Graduation in Horticulture from GBPUA&T, Pantnagar and Ph.D. in Horticulture from BHU, Varanasi.

Dr. Singh has served Tilak Dhari Post Graduate College in various capacities like In-charge Garden Section, Captain in NCC, Member of the Sports Committee, Academic Counselor of IGNOU study center and Former Proctor of the college. Dr. Singh has 19 years teaching experience of horticulture. Dr. Singh has more than 62 publications to his credit (research papers 25, proceeding papers 5, conference/seminar papers 10, books 2, book chapters 10 and popular/semi-scientific articles 10). He had been chairman of the organizing committee of One National Workshop. Dr. Singh qualified ICAR NET in Horticulture and Vegetable Science and received Senior Research Fellowship in ICFRE during Ph.D. programme. He is fellow of Indian Society of Vegetable Science and Indian Society of Horticulture Science. He has supervised 4 M.Sc. (Ag.) and 3 Ph.D. students as Chairman of Advisory Committee.

Dr. Bijendra Kumar Singh is presently working as Assistant Professor, Department of Horticulture, Tilak Dhari Post Graduate College, Jaunpur (U.P.). He did his Graduation in Agriculture from VBSPU, Jaunpur, Post Graduation in Horticulture from BBAU, Lucknow and Ph.D. in Horticulture from BHU, Varanasi. He has specialization in Pomology.

Dr. Singh has served Tilak Dhari Post Graduate College in various capacities like Member of Campus Greenery of college under IQAC committee and Academic Counselor of IGNOU study center. Dr. Singh has 143 publications (research papers 30, proceeding papers 3, conference/seminar papers 50, book chapters 25 and popular/semi-scientific articles 35). He had been convener of the organizing committee of One National Workshop. Dr. Singh qualified ICAR NET in Fruit Science and received University Grant Commission (UGC) fellowship during his Ph.D. He is a life member of renowned societies and journals in India. Asian PGPR Society of Sustainable Agriculture, International College of Nutrition's, Society for Advancement of Research on Pomegranate, Trends in Biotechnology & Biological Sciences, Advances in Life Sciences and Trends in Biosciences. He has supervised 4 M.Sc. (Ag.) and 3 Ph.D. students as Chairman and Member of Advisory Committee.

Basic Horticulture and Fruit Production Technology

Rajaneesh Singh
Head
Department of Horticulture
Tilak Dhari Post Graduate College
Jaunpur, Uttar Pradcsh

Bijendra Kumar Singh
Assistant Professor
Department of Horticulture
Tilak Dhari Post Graduate College
Jaunpur, Uttar Pradesh

nipa®

NIPA® GENX ELECTRONIC RESOURCES & SOLUTIONS P. LTD.
New Delhi-110 034

**NIPA® GENX ELECTRONIC
RESOURCES & SOLUTIONS P. LTD.**

101, Vikas Surya Plaza, CU Block, LSC Market
Pitam Pura, New Delhi 110 034, India
Phone: + 91 (11)27 34 17 17 Fax: + 91(11) 27 34 16 16
Email: info@nipabooks.com
Web: www.nipabooks.com

Feedback at feedbacks@nipabooks.com

ISBN : 978-93-95763-94-3

Composed and Designed by NIPA

World Noni Research Foundation

64, Third Cross Street, Second Main Road, Gandhi Nagar, Adyar, Chennai – 600 020
Telefax: 044-2442 3601 Email: mail@worldnoni.org Website: www.worldnoni.org

Dr. Kirti Singh
FNASc, FNAAS, FNABS
CHAIRPERSON

Foreword

Fruit growing is one of the important and age old practices. Practiced in India since ancient times, cultivation of fruit crops plays an important role in improving the overall status of mankind. The standard of living of people of a country is depending upon the production and per capita consumption of fruits. Fruit crops also play an important role in the human diet because of their ingredients. Fruits work as protective food against a number of disease and ailments. There is a great potential for export of fruits from India. However, it can be achieved by improving the quality of fruits through judicious management and cultivation practices. Horticulture production in India has now surpassed the cereal production.

The present text book entitled "Basic Horticulture and Fruit Production Technology" deals with basic elementary information which is essential at undergraduate and postgraduate levels for the students of horticulture. This book fulfills the gap and provides all aspects of the basic horticulture and fruit production technology. It covers fundamental aspects of science of horticulture which are lacking in many books available on the subject.

In my opinion, this text book shall be extremely helpful, to the students of horticulture in various colleges and universities as well as to the teachers engaged in teaching the subject. I have great pleasure to complement and congratulate the authors for their enthusiasm and hard work in bringing out useful and informative publication.

Date : 05-09-2019
Place : Jaunpur

(Kirti Singh)
Former Vice-Chancellor

Preface

Fruits are the nature's gift to mankind. They are not only delicious and refreshing but are also the chief source of vitamins, minerals and proteins. It may not be exaggeration to say that 80-90 per cent of problems arising from under nutrition or malnutrition can be well managed with indigenous medicines and fruits are an important part of these.

The book is primarily meant for the student of graduate and postgraduate in the field of horticulture of all agricultural universities in India and neighbouring countries. The information included in this book is considered to be of utmost value to student of horticulture fruit growers, nursery man, gardeners, subject matter specialist and other person's engaged in the field of horticulture. In this edition I have shared my personal experience on fruit crops acquired during last nineteen years.

The book will proved boon for student contesting Indian Council of Agricultural Research Junior and Senior Fellowship, Agriculture Research Services examination and NET in the field of horticulture. I am highly thankful to management of the college and principal for providing me always moral support in writing of this book. We have great pleasure in acknowledging moral support, encouragement valuable suggestion and connective criticism received from Dr. Hari Baksh, Mr. Raj Pandey and all agriculture faculty members, T.D. College.

I am highly grateful to Prof. Anil K. Singh (BHU), Prof. Ranveer Singh, Prof. J.P. Tiwari and Prof. Harihar Ram who always encouraged me for preparing this manuscript.

I feel immense pleasure to express our heartfelt gratitude to my elder brother Prof. I.S. Singh, Mr. Avaneesh Singh and all family members for their constant inspiration, guidance, encouragement and blessing.

At the last but not the least I can never forget the co-operation rendered by my wife Smt. Priya Singh and my children Pranjal Singh and Kishlay Singh without whom this task would not have been completed.

Jaunpur **(Bijendra Kumar Singh)** **(Rajaneesh Singh)**

Contents

Section I: Basic Horticulture

Section II: Fruit Production Technology

Section I: Basic Horticulture

1

Introduction

- The term **Horticulture** is derived from two **Latin words** ***Hortus*** meaning = **Garden** and ***Cultura*** meaning = **Cultivation**.
- In ancient days the gardens had protected enclosures with high walls or similar structures surrounding the houses.
- The enclosed places were used to grow fruit, vegetables, flowers and ornamental plants. Therefore, in original sense "Horticulture refers to cultivation of garden plant within protected enclosures".

Definition: Horticulture is the branch of science which deals the cultivation, utilization and management of horticultural crops eg. fruits, vegetables, flowers, spices, plantations, medicinal and aromatic plants.

Branches of Horticulture

- Horticulture is a wide field which includes a great variety and diversity of crops.
- The science of horticulture can be divided into several branches depending upon the crops it deals with.
- Following are the branches of horticulture.

i. **Pomology:** Study of fruit crops.

ii. **Olericulture:** Cultivation of vegetables.

iii. **Floriculture:** Cultivation of flower crops.

iv. **Plantation crops:** Cultivation of coconut, arecanut, rubber, coffee, tea, etc.

v. **Spices crops:** Cultivation of cardamom, pepper, nutmeg etc.

vi. **Medicinal and aromatic crops:** Cultivation of medicinal and aromatic crops.

vii. **Post harvest technology:** Which deals with post harvest handling, grading, packaging, storage processing, value addition, marketing etc, of horticulture crops?

A. Fruit Crops

1. India is the second largest producer of fruits after Brazil.
2. A large variety of fruit crops are grown in India. Of these, mango, banana, citrus, papaya, guava, pineapple, sapota, jackfruit, litchi, grapes, apple, pear, peach, plum, walnut etc. are the important ones.
3. India accounts for 10 per cent of the total world production of fruits.
4. It leads the world in the production of mango, banana, sapota and acid lime besides recording highest productivity in grape.
5. The leading fruit growing states are Maharashtra, Karnataka, Andhra Pradesh, Bihar and Uttar Pradesh.

B. Vegetable Crops

6. More than 40 vegetables belonging to solanaceaeous, cucurbitaceous, leguminous, cruciferous, root crops and leafy vegetables are grown in Indian tropical, sub-tropical and temperate regions.
7. Important vegetables grown in India are onion, tomato, potato, brinjal, peas, beans, okra, chilli, cabbage, cauliflower, bottle gourd, cucumber, watermelon, carrot, radish etc.
8. India ranks second in vegetable production next to China.
9. India occupies first position in cauliflower, second in onion, third in cabbage in the world.
10. West Bengal, Orissa, Uttar Pradesh, Bihar, Maharashtra, Karnataka are the important states for horticultural crop production.

Table 1 : Differences between fruits and vegetables

S.No.	Fruits	Vegetables
1.	Fruit plants are perennial in nature	Most of the vegetables are annuals
2.	Fruit plant are sexually and asexually propagated	Mostly majority of them are sexually propagated
3.	Fruit plants require special practices like training and pruning and are required seasonally.	Cultivation of vegetable is seasonal and special techniques like pruning and training are generally not required
4.	Fruit plants are generally woody in nature	Vegetable plants are generally non-woody
5.	Only fruit is edible but sometimes false fruit also edible (eg. Fleshy thalamus of apple)	All parts of the plant are edible
6.	Mostly consumed raw after ripening	Generally consumed after cooking
7.	Fruits are always acidic in nature	Vegetables are alkaline in nature

C. Floriculture

11. In India, flower cultivation is being practiced since ages.
12. It is an important/integral part of socio-cultural and religious life of Indian people.
13. It has taken a shape of industry in recent years.
14. India is known for growing traditional flowers such as jasmine, marigold, chrysanthemum, tuberose, crossandra, aster, etc.
15. Commercial cultivation of cut flowers like, rose, orchids, gladiolus, carnation, anthurium, gerbera is also being done.
16. The important flower growing states are Tamil Nadu, Karnataka, Andhra Pradesh, Maharashtra, West Bengal, Sikkim, Jammu & Kashmir, Meghalaya, etc.

D. Plantation Crops

17. The major plantation crops include coconut, arecanut, oil palm, cashew, tea coffee, rubber cocoa, betel vine, vanilla etc.
18. The leading states are Karnataka, Kerala, Tamil Nadu, Andhra Pradesh, Maharashtra, Goa, Assam etc.

E. Spices

19. They constitute an important group of horticulture crops and are defined as vegetable products or mixture thereof.
20. Free from extraneous matter used for flavouring, seasoning and imparting aroma in foods.
21. India is known as **home of spices** producing a wide variety of spices like black pepper, cardamom, ginger, turmeric, chilli, coriander etc.
22. Major spice producing states are Kerala, Andhra Pradesh, Gujarat, Rajasthan, Maharashtra, Karnataka, Orissa, Tamil Nadu etc.

F. Medicinal and Aromatic Plants

23. India has diverse collection of medicinal and aromatic plants species distributed throughout the country.
24. It has more than 9,500 species with medicinal properties.
25. Demand for these crops is increasing progressively in both domestic and export markets.
26. Important medicinal plants are Isabgol, Senna, Opium poppy, Periwinkle, Coleus, Ashwagandha, etc. and aromatic plants are Japanese mint, Lemon grass, Citronella, Davana, Patchouli etc.

2

Importance and Scope of Horticulture

1. Total horticulture covers an area of 27.1 m ha and production 305.8 m ton in the country annually (Indian Horticulture Database, 2018).
2. Fruit crops cover an area of 8.4 m ha and vegetable crops 12.2 m ha. Accordingly, 98.7 m ton of fruits and 182.1 m ton of vegetables are produced in the country annually (Indian Horticulture Database, 2018).
3. Apart from fruits and vegetables, floriculture industry in India comprising of florist trade, nursery plants, potted plants, seed and bulb products is being observed as sunrise industry.
4. The developed flower market in the country during 2018 is with area of 0.32 m ha with a production of 2.7 m ton flowers.
5. The traditional flowers are grown on a large area on a commercial scale. These flowers are mostly grown for loose flower purpose.
6. Area under cut flowers like rose, chrysanthemum, gladiolus, carnation and orchids is increasing day by day.
7. Plantation crops are another potential sector with lot of opportunities for employment generation, foreign exchange earnings and overall supporting livelihood sustenance of mankind at large.

Horticulture is important due to the following considerations

1. As a source of variability in produce.
2. As a source of nutrients, vitamins, minerals, flavour, aroma, alkaloids, oleoresins, fibre, etc.
3. As a source of medicine.
4. As an economic proposition as they give higher returns per unit area in terms of energy, money, job, etc.

5. Employment generation fruit crops requires 860 man days/annum as against 143 man days/annum for cereal crops where as the crops like grapes, banana and pineapple needs 1000- 2500 man days per annum.
6. Effective utilization of waste land through cultivation of hardy fruits and medicinal plants.
7. As a substitute for family income being the component of home garden/ kitchen garden.
8. As a foreign exchange earner, has higher share compare to agriculture crops.
9. As an input for industry being amenable to processing, especially fruit and vegetable preservation industry.
10. Aesthetic consideration and protection of the environment.
11. Religious significance in the country.

In short and sweet horticulture supplies quality food for health and mind, more calories per unit area, develops better resources and yields higher returns per unit area. It also enhances land value and creates better purchasing power for those who are engaged in this industry. Therefore, *horticulture is important for health, wealth, hygiene and happiness.*

Scope of Horticulture

Like any other things, scope of horticulture depends on incentive it has for the farmers, adaptability of the crops, necessity and facilities for future growth through inputs availability and infrastructure for the distribution of produce/ marketing etc.

1. Incentive for the farmer

- The biggest incentive for the farmer is money.
- Horticultural crops provide more returns in terms of per unit area of production, export value, value addition compared to agricultural crops.

2. Adaptability

- India is bestowed with a great variety of climatic and edaphic conditions as we have climates varying from tropical, subtropical, temperate and within these humid, semi-arid, arid, frost free temperate etc.
- Likewise we have soils from loam, alluvial, laterite, medium black, rocky shallow, heavy black, sandy etc., and thus a large number of crops can be accommodated with very high level of adaptability. Thus, there is lot of scope for horticultural crops.

3. Necessity

- After having achieved the self sufficiency in food, nutritional security for the people of the country has become the point of consideration/priority.
- To meet the nutritional requirement in terms of vitamins and minerals horticulture crops are to be grown in sufficient quantities to provide a bare minimum of 120 g of fruits and 300 g of vegetables per head per day with a population of above 120 crores.
- Good land is under pressure for stable food, industry, housing, roads and infrastructure due to population explosion and only wasteland had to be efficiently utilized where cultivation of annuals is a gamble due to restricted root zone and their susceptibility of abiotic stress. These lands can be best utilized to cultivate hardy horticultural crops like fruits and medicinal plants.
- At present our share in international trade of horticultural commodities is less than one per cent of total trade. Moreover, these commodities (spices, coffee, tea etc.,) fetch 10-20 times more foreign exchange per unit weight than cereals and therefore, taking advantage of globalization of trade, nearness of big market and the size of production, our country should greatly involve in international trade which would provide scope for growth.

4. Export value

- Among fresh fruits-mangoes and grapes; in vegetables- onion and potato; among flowers, roses; among plantation - cashewnut, tea , coffee, coconut, arecanut, and spice crops like black pepper, cardamom, ginger, turmeric, chillies, etc., constitute the bulk of the export basket.
- *European* and *gulf* countries are major importer of horticultural produce.

Table 2 : Major countries for export of horticultural produce from India

S. No.	Commodity	Major importer	Share value (%)
1	Fruits and vegetables	United Arab Emirates	28.00
2	Flower crops	U.S.A	37.70
3	Processed products	Soviet Union Countries	14.60
4	Spices	U.S.A	43.50
5	Cashew	U.S.A	40.00

- In the recent past communication and transport system have improved, investment in food industry has increased which will support growth of horticulture through quick deliverance and avoidance of waste.

Reasons for scope of horticulture in India are

1. To exploit the great variability of agro climatic conditions in the country.
2. To meet the need for fruits, vegetables, flowers, spices, beverages in relation to population growth based on minimum nutritional security and for other needs.
3. To meet the requirement of processing industry.
4. To substitute import and increase export.
5. To improve the economic conditions of the farmers and to engage more labourers to avoid the problem of unemployment.
6. To protect environment.

Other importance

1. Similar to forest trees these horticultural trees will maintain the ecosphere.
2. They help in transforming the micro climate.
3. Provides shelter to birds, reptiles and other micro organisms and add to the geo-ecological diversity on the land.
4. Provides thrust to the writers, poets, thinkers and analysts there by keeps their cultural impulse alive.
5. Adds to the survival of life-spheres of living entity.

Horticultural crops and human nutrition

1. Fruits and vegetables play an important role in balanced diet.
2. These provide not only energy rich food but also provide vital protective nutrients/elements and vitamins.
3. Comparatively fruits and vegetables are the cheapest source of natural nutritive foods.
4. Since most of Indians are vegetarians, the incorporation of horticulture produce in daily diet is essential for good health.
5. Realizing the worth of fruits and vegetables in human health, Indian Council of Medical Research (ICMR) recommended the use of 120g fruits and 300g vegetables per capita per day.
6. With the growing awareness and inclination towards vegetarianism worldwide the horticulture crops are gaining tremendous importance.

Functions of fruits and vegetables in human body

1. Fruits and vegetables provide palatability/taste,
2. Improves appetite and provides fiber to overcome constipation.

3. They neutralize the acids produced during digestion of proteins and fatty acids.
4. They improve the general immunity of human body against diseases, deficiencies etc.
5. They are the important source of vitamins and minerals for used in several bio-chemical reactions occur in body.
6. Fruits and vegetables provide higher energy value per unit area compared to cereals.
   - Fruits are also a good source of energy eg. Avocado, Olive etc.,
   - Fruits are also a good source of enzymes which are helpful in metabolic activities leading to proper digestion of food eg. Jamun and Papaya.
   - All fruits have one or the other medicinal value.
   - They should be eaten in adequate quantity.
   - Regular consumption of fruits reduces obesity, maintain health and increase the longevity of life.
   - Fruits are attractive in appearance, delicious in taste and easily digestible. Therefore, they are liked by young and old alike.

Importance of fruit in human diet

- For balance diet fruit play an important role in health, happiness and prosperity.
- Fruits are the rich source of mineral and vitamins without which human cannot maintain their proper health and resistance to diseases.
- Deficiency of mineral and vitamins causes disturbances of metabolism and resulting ill health beside these protein and cellulose are found in the fruit with stimulate the intestinal activity and protect the human body against various type of disorder.
- Man cannot survive bread alone therefore at least 120 gm fruit capita per day is required in addition to pulse, cereals, milk, vegetable etc.
- For keeping the body system fit, use of fruit and vegetable have become must in our daily diet, especially in present era of environment pollution.

Table 3 : Some of the essential nutrients provided by different fruits are:

Vitamins/Minerals	Role in human body	Sources
Vitamin-A	1. Essential for growth and reproduction. 2. Helps in resistance to infections, increases longevity and decreases senility. 3. Deficiency causes, night blindness, xeropthalmia, retardation in growth, roughness in skin, formation of stones in kidney.	*Mango, Papaya, Persimon, Dates, Jack fruit, Walnut, Oranges, Passion fruit, Loquât etc.*
Vitamin-B_1	1. For maintaining good appetite and normal digestion. 2. Necessary for growth, fertility, lactation and for normal functioning of nervous system. 3. Deficiency causes beri-beri, paralysis, loss the sensitivity of skin, enlargement of heart, loss of appetite and fall in body temperature.	*Walnut, Apricot, Apple, Banana, Grapefruit, Plum and Almond*
Vitamin-B_2	1. Important for growth, health of skin and for respiration in poorly vascularised tissue such as the cornea. 2. Deficiency causes pellagra and alopecia, loss of appetite, loss of weight, sore throat, and development of cataract, swollen nose and baldness.	*Bael, Papaya, Litchi, Pomegranate, Wood apple and Pineapple.*
Vitamin-C	1. Deficiency causes scurvy, pain in joints, swelling of limbs, unhealthy gums, tooth decay, delay in wound healing and rheumatism.	*Barbados cherry, Aonla, Guava, Lime, Lemon, Sweet oranges, Ber, Pineapple and Pear.*
Fat	Walnut, Almond, Avocado	
Fibre	Guava, Pomegranate, Aonla, Grape, Amaranth, Mustard, Beet leaf, Spinach etc.	

Table 4 : Minerals are essential for the growth and development for the human body:

Minerals	Deficiency causes	Sources
Calcium	Causes Rickets, Osteomalacia.	*Sitaphal, Ramphal, Fig, Phalsa, Citrus, Sapota, Grapes, West Indian Cherry etc.*
Phosphorous	Essential for cell multiplication of bones and soft tissues. Helps in liberation of energy on oxidation of carbohydrates.	*Wood apple, Avocado, Dates, Pomegranate and Grape raisins.*
Proteins	Important for body growth, formation and maintenance of body tissues	*West Indian cherry, Avocado, Custrad Apple, Banana, Apricot, Guava, Grapes etc.,*
Iron	Act as oxygen carrier in the body.	*Karonda, Date palm, Grape raisins, West Indian Cherry, Guava, Sitaphal, Avocado, Sapota, plum etc.*

Importance and Scope of Fruit Crops

Fruit growing is one of the important and age old practices, in India since ancient times. Cultivation of fruit crops plays an important role in overall status of the mankind and the nation. The standard of living of the people of a country is depending upon the production and per capita consumption of fruits. Fruit growing have more economic advantages.

1. Economic importance

- **High productivity:** High yield per unit area: From a unit area of land more yield is realized from fruit crops than any of the agronomic crops. The average yields of Papaya, Banana and Grapes are 10 to 15 times than that of agronomic crops.
- **High net profit:** Through, the initial cost of establishment of an orchard is high it is compensated by higher net profit due to higher productivity or high value of produce eg- Grapes/Mango/Banana-20-40t/ha-1.5-2.5 lakh/ha.
- **Source of raw material for agro based industries:** Fruit farming provides raw materials for various agro based industries- canning and preservation (fresh fruits), coir industries (coconut husk), pharmaceutical industry (Aonla, Papaya, Jamun) Transporting and packaging industries etc.
- **Efficient utilization of resources:** Growing of fruits being perennial in nature, enables grower to remain engaged throughout the year in farm operations and to utilize fully the resources & assets like machinery, labour, land water for production purpose throughout the year compared to agronomic crops.
- **Utilization of waste and barren lands for production:** Although, most of the fruits crops require perennial irrigation and good soil for production, there are many fruit crops of hardy in nature, Mango, Ber, Cashew, Custard apple, Aonla, Phalsa, Jamun etc. which are grown on poor shallow, undulated soils considered unsuitable for growing grain/ agronomical crops.
- **Foreign exchange:** Many fresh fruits, processed products and spices are exported to several countries earning good amount of foreign exchange.

2. Nutritional importance

Importance of fruits in human diet is well recognized. Man cannot live on cereals alone. Fruits and vegetables are essential for balanced diet and good health. Nutritionist advocates 120 gm of fruits and 300 gm. vegetables per capita per day in addition to cereals, pulses, egg etc. fruits and vegetables are good sources of vitamins and minerals without which human body cannot maintain proper health and develop resistance to disease they also contain pectin, cellulose, fats, proteins etc.

Fruits- as sources of vitamins

1. Vitamin-A- Mango, Papaya, Jackfruit, Banana, Dates.
2. Vitamin-B- Cashew nut, Almond, Banana, Apple, Bale, Litchi, Papaya and Pomegranate.
3. Vitamin-C- Aonla, Citrus fruits, Pineapple, Ber, Guava, Strawberry, Tamarind etc.

Fruits as a source of minerals: Such as Ca, Fe, P- the fruits are- Almond, Cashew, Guava, Jamun, Fig, Karonda, and Mango.

Digestive enzymes- Papaya- papaine-9, proteolitic enzyme.

Fruits have medicinal value: The fruits like Aonla Pomegranate, Kokum, Jamun, Bael, Ber etc., have great medicinal value,

1. Papaya reduces night blindness,
2. Citrus juice reduces acute dihorrhea.
3. Aonla triphala (chawan prash) - digestion.
4. Jack fruit (Jackoline) - prevents Aids.

Other importance: fruit growing in kitchen gardens helps to reduce family budget on purchase of fruits.

- Planting of fruits trees, maintains ecological balance and to increase precipitation of the locality.
- Fruit tree farming also reduces soil erosion, silting and air pollution.
- Generate employment being highly intensive & skillful enterprise generates employment even for trained persons.

3

Horticultural Zone of India

The Indian subcontinent is bestowed with a great variety of climate and soil conditions. Broadly the country can be divided into tropical, subtropical and temperate regions, within each broad category there are differences due to rainfall, humidity, altitude etc. Considering these aspects six different horticultural zones have been identified so that appropriate choice of the crops can be made and development is planned. They are;

Temperate: Kashmir, Himachal Pradesh, North Uttarakhand, Sikkim and part of Arunachal Pradesh.

N.W. Subtropical: Punjab, Haryana, Rajasthan, Central Uttar Pradesh and North Madhya Pradesh.

N.E. Subtropical: Bihar, Jharkhand, Assam, Meghalaya, Nagaland, Manipur.

Central tropical: South Madhya Pradesh, Chhattisgarh, Gujarat, Maharashtra, Orissa and West Bengal.

Southern tropical: Karnataka, Andhra Pradesh and Tamil Nadu.

Coastal tropical humid: Konkan, Goa, Kerala, Western Ghats, Eastern Ghats in Tamil Nadu, Andhra Pradesh and Orissa.

- To exploit the potential of a crop and its sustenance, right choice based on climate and soil is necessary otherwise the management of the crop becomes difficult and the cost of cultivation increases. To be precise, most adaptable crop should be chosen for nourishment.

Table 5: Climatic requirements for important fruits of India

Mango	Tropical and sub tropical.
Citrus	Subtropical but can be grown under temperate conditions.
Grape	Temperate but can be grown under subtropical and tropical conditions.
Peachs	Temperate but low chilling varieties can be grown under subtropical conditions.
Sapota	Tropical but can be grown under subtropical conditions which are free from frost.
Papaya	Tropical and mild subtropical climate.
Banana	Tropical, can be grown under subtropical climate provided it is free from hot winds and frost.
Almond	Temperate but some low chilling varieties can be grown under subtropical climate.
Apple	Temperate but low chilling varieties can also be grown on lower hills.

- Climate is one of the important complex factors which influence the fruit production.
- Atmospheric conditions include rainfall, humidity, sunshine, wind and other factors.
- The fruit growing zones are classified based on the climate factors.

Fruit zones of India

Climate is one of the important complex factors which influences the fruit production is defined as the general temperature and atmospheric conditions of an area, over an extended period of time. Atmospheric conditions include rainfall, humidity, sunshine, wind and other factors. The fruit growing zones are classified based on the climate is factors.

Classified fruit zones

1. **Tropical fruit zone:** This class includes fruit crops which are ever green unable to endure cool temperature but can tolerate warm temperature of about 100^{0}F. The fruit plants of this zone need strong sunshine warm and humid climate and a very mild winter. They cannot stand against frost. Area under this zone includes West Bengal, Parts of Punjab, Haryana, Rajasthan Orissa, Maharashtra, Andhra Pradesh, Karnataka, Tamil Nadu and Kerala. Fruits crops: Banana, Pineapple, Sapota, Papaya, Cashew, Pomegranate.
2. **Sub-tropical fruit zone:** This class includes fruit crops intermediate characters to tropical and temperatures. The summer is hot and dry and winter is less mild. They may be either deciduous or ever green & are usually able to withstand a low temperature but not the frost. Some require chilling for flower bud differentiation the fruits grow mostly in plains, the fruits includes, Citrus, Grapes, Phalsa, Fig, Guava, Pomegranate, Banana etc. This fruit zone covers the plains of Punjab, Uttar Pradesh, Parts of Bihar, Madhya Pradesh, West Bengal, Maharashtra, Rajasthan, Karnataka, Andhra Pradesh, Tamil Nadu, Kerala, Orissa. etc.
3. **Temperate fruit zone:** This class of fruits grows successfully in cold regions where temperature falls below freezing point during winter. During the cold season, the trees shed their leaves and go into rest period. For breaking the rest/dormant period, a definite chilling period is required. This chilling temperature helps the plants to put forth new growth, flowering and fruiting with the onset of spring season. The regions under this zone are Jammu & Kashmir, Kuluvally, Himachal Pradesh, Parts, Peaches, Plum, Cherries, Almond, Walnut, Strawberry, Apricot, persimmon, Pecan nut, Kiwi fruit etc.

4. **Arid zone:** The arid zone has an extreme climatic condition, high temperature low humidity, rainfall is very low and its distribution is erratic, poor textured soil. The area of Rajasthan (62%) and Gujarat (20%) parts of the Punjab, Haryana, Karnataka & Maharashtra the crops are Phalsa, Date palm, Pomegranate, Ber, Custard apple, Tamarind etc.
5. **Semi-arid zone:** This region exhibits low and erratic rainfall, low humidity and high temperature fruits of arid region can be cultivated in this zone also Mango, Sapota, Guava, Jack, Avocado, Ber, Pomegranate and Tamarind etc.
6. **North-Eastern sub-tropical zone:** All tropical and sub-tropical fruits are grown in this region. The parts are Bihar, Assam, Meghalaya, Manipur, Parts of West Bengal and Uttar Pradesh etc.
7. **North-Western region:** It is again classified in to 4 temperate- Low winter temperature, Dry temperature-highly cold condition, Sub- Temperate- winter temperature & lesser cold, Low hill valley- low winter temperature & lesser cold. Parts of Jammu & Kashmir, Himachal Pradesh, hills of Uttar Pradesh, South of Punjab, Haryana.
8. **Central tropical fruit zone:** This region covers Southern parts of Madhya Pradesh, Maharashtra Orissa, parts Andhra Pradesh, West Bengal, Gujarat etc.
9. **South tropical fruit zone:** Karnataka, Tamil Nadu, Kerala & Andhra Pradesh.
10. **Coastal tropical fruit zone**: Kerala, Goa, Diu-Daman, Tripura, Coastal parts of Maharashtra, Andhra Pradesh, West Bengal, Tamil Nadu, Orissa, Karnataka.
11. **Humid zone fruit crops:** This region is characterized by low temperature and high humidity. The crops are Litchi, Strawberry, Avocado, Mangosteen, Passion fruit etc. Apart from these fruit zones, India has been classified in to 21 agro ecological regions based on the physiographic of soils, bioclimatic types and growing periods.

4

Classification of Fruit Plants

Classification of fruits: Classification is the system of grouping or placing of an individual according to nomenclature. It is very useful to the pomologist, it helps to:

- To identify and naming the crop.
- To study the close relationship.
- To know their hybrids and crossing behavior.
- To know their compatibility & inter grafting ability.
- To know their adoptability to soil & climate.

Pomology: Pomology is a branch of horticulture which deals the cultivation utilization and management of the fruit crops. eg. Mango, Banana, Citrus, Papaya, Guava etc. The term pomology is a combination of two words "Poma" in Greek means Fruits later subsequently transfer in to "Pome" in Latin word means Fruits and Logos means Study.

- **Basic pomology:** Study of basic aspects of fruit production like pruning and training, water management, use of PGRs.
- **Commercial pomology:** It is concerned with commercial production of fruits.
- **Systematic pomology:** It may be concerned with classification and nomenclature like kingdom, order, class, genus and species. These terms are given by DeCondole.

A. Classification of fruits based on climate adaptability

In this classification, the fruits trees are categorized into three recognized groups.

i. **Temperate fruits**: Temperate fruit plants are exacting in their climate requirement. They are grown only in place where winter is distinctly cold, require as exposure of specific chilling temperature for certain period without which they do not flower. These fruit plants are generally deciduous and stand frost. e.g., apple, almond, peach, pear, plum, strawberry, apricot, persimmon, cherimaya, walnut, peanut, hazel nut, cherry, pistachio nut and kiwifruits etc.

ii. **Tropical fruits**: Tropical fruit plants are generally evergreen and are extremely sensitive to cold. The plants are generally grown in climatic conditions prevailing in the region between the tropic of Cancer (23^0(27) N latitude) and the tropic of Capricorn (23^0(27) S latitude). They do well under lesser fluctuations of diurnal temperature, light and dark periods they require a moist warm climate but are capable of withstanding dry weather in some cases e.g., mango, banana, papaya, sapota, etc.

iii. **Sub-tropical fruits:** The fruit crops grown under a climatic condition between temperate and the tropical are known as subtropical fruit crops. They may be either deciduous or evergreen and are usually able to withstand a low temperature but not the frost. They are also quite adoptive to fluctuations of light and dark period during day and night. Some subtropical fruit plants require chilling for flower bud differentiation. e.g., grape, citrus, durian, jackfruit, etc.

B. Classification based on bearing habit

On the basis of bearing habit, fruit trees are classified in to six categories to facilitate cultural operation like pruning, skiffing, heading back etc.

a. Fruit buds bore terminally and giving rise to inflorescence without leaves e.g., Mango, Cherry, etc.

b. Fruit buds borne terminally and unfolding to produce leafy shoots which terminate in flower clusters. e.g., Apple.

c. Fruit buds borne terminally and unfolding to produce leafy shoots with flower or flower clusters e.g., Guava.

d. Fruit bud borne laterally containing flower parts only and giving rise to inflorescence without leaves or leaves present, they are reduced in size., e.g., Citrus.

e. Fruit bud borne laterally and unfolding to produce leafy shoots terminally in flower clusters this type of flowering is noticed in Grapes and Cashewnut.

f. Fruit buds borne laterally and unfolding to produce leafy shoots with flower clusters in leafy axils eg. Fig.

C. Based on fruit morphology

1. **Simple fruit-**

   Berry : Banana, Papaya, Grape, Sapota, Avocado

2. **Modified berry-**

   Balusta : Pomegranate

   Amphisarca : Woodapple, Bael

Pepo : Water melon
Pome : Apple, Pear, Laquat
Drupe (Stone) : Mango, Pear, Plum
Hesperidium : Citrus
Nut : Cashew, Litchi, Walnut, Rambutarn
Capsule : Anola, Carambola

3. **Aggregate fruits-**

Etario of berries : Custard apple, Raspberry

4. **Multiple fruit-**

Syconus : Fig
Sorosis : Jackfruit, Pineapple, Mulberry

D. Based on rate of respiration

Climacteric	Non-climacteric
Mango, Banana, Sapota, Guava,Papaya, Apple, Fig, Peach, Pear, Plum, Annona, Tomato	Citrus, Grape, Pomegranate, Pineapple, Litchi, Ber, Jamun, Cashew, Cucumber, Cherry, Strawberry.

(Climacteric fruits produce much larger amount of ethylene than non climacteric fruits)

E. Based on photoperiodic responses

Long day	Short day	Day neutral plant
Passionfruit, Banana, Apple	Strawberry, Pineapple, Coffee	Papaya, Guava

F. Based on relative salt tolerance

Highly tolerant	Medium tolerant	Highly sensitive
Datepalm, Ber, Aonla, Guava, Coconut, Khirni	Pomegranate, Cashew, Fig, Jamun, Phalsa	Mango, Apple, Citrus,Pear, Straw berry

G. Based on relative acid tolerance

Highly tolerant	Medium tolerant	Highly sensitive
Stawberry, Raspberry, Fig, Bael, Plum	Pineapple, Avocado, Litchi	-

H. Based on longevity

a. Very Long longevity - >100yrs- Datepalm, Coconut, Arecanut
b. Long longevity - 50-100yrs - Mango, Tamarind
c. Medium - 10-50yrs - Litchi, Guava, Pomogranate
d. Short - Pineapple, Banana

Botanical classification based on botanical relationship with genomes.

Family	Crops	Scientific Name	Fruit Type	Chromosome No.	Origin
Monocotyledoneae:					
Musaceae	Banana	*Musa paradisiaca, Musa balbisiana*	Berry	22,33,44	Indo-Malayan
	Plantain	*Musa acuminata*	Berry	22,33,44	Indo-Burma
Bromeliaceae	Pineapple	*Ananas comosus*	Sorosis	50,75,100	Brazil
Palmae/ Aracaceae	Peach plum	*Guillielma gasipes*	Drupe		West Asia
	Date palm	*Phoenix dactylifera*	Drupe	36	West Asia
Dicotyledoneae:					
Anacardiaceae	Mango	*Mangifera indica*	Drupe	40	South East-Asia
	Pistachionut	*Pistachia vera*	Nut	30	Iran/Iraq
	Cashewnut	*Anacardium occidentalle*	Nut	42	Brazil
	Indian hog plum	*Spondias pinnata*	Drupe		
	Hog plum	*Spondias cytherea*	Drupe		
Annonaceae	Custard apple	*Annona squamosa*	Aggregate berries	14	West Indies
	Bullock heart	*Annona reticulata*	Aggregate berries	14	
	Sour sap	*Annona muricata*	Aggregate berries	14	South America
	Cherimoya	*Annona cherimoya*	Aggregate berries	14	Bovilia
	Atemoya	*Annona atemoya*	Aggregate berries	14	Man made hybrid
	Pond apple	*Annona glebra*	Aggregate berries	14	
Apocyanaceae	Karonda	*Carissa carandus*	Berry	22	
Caricaceae	Papaya	*Carica papaya*	Berry	18	Tropical America
Ebenaceae	American persimmon	*Diospyros vixgiana*	Berry		China
	Persimmon	*Diospyros kaki*	Berry	90 (6x)	China
Euphorbiaceae	Aonla/Nelli	*Emblica officinalis*	Capsule (Drupe)	28	Madagascar
	Star gooseberry	*Phyllanthus acidus*	Berry		
Guttiferae/ clusiaceae	Cowphal	*Garcinia cowa*	Berry		

	Malabar tamarind	*Garcinia cambogia*	Berry		
	Mangosteen	*Garcinia mangostana*	Berry	28	Malayan Archipelago
Lauraceae	Avocado	*Persea americana*	Berry	24	Central America
Malphighiaceae	Barbados cherry/ West Indian cherry	*Malphigia puncifolia*	Drupe	40	Trinidad and Tobago
Moraceae	Aini	*Artocarpus hirsuta*	Sorosis	56	
	Jack fruit	*Atrocarpus heterophyllus*	Sorosis	56	India
	Bread fruit	*Artocarpus altilis*	Sorosis	56	Indo-Malayan
	Monkey jack	*Artocarpus lakoocha*	Sorosis	56	Western Ghats
	Fig	*Ficus carica*	Synconus	26	
	Mulberry	*Morus alba*	Sorosis	308	
Myrtaceae	Guava	*Pisidium guajava*	Berry	22	Tropical America
	Jamun	*Syzygium cuminii*	Drupe	40	India
	Rose apple	*Syzygium jambos*	Drupe	20	
Oleaceae	Indian olive	*Olea ferruginea*	Drupe		
	Olive	*Olea europaea*	Drupe	46	Mediter-ranean region
Oxalidaceae	Bilimbi	*Averrhoa bilimbi*	Berry	24	
	Carambola	*Averrhoa carambola*	Berry	24	Sri Lanka
Passifloraceae	Passion fruit	*Passiflora edulis*	Berry	18	Brazil
Proteaceae	Macadamia nut	*Macadamia ternifolia*	Nut	48	Australia (East)
Punicaceae	Pomegranate	*Punica granatum*	Balusta	18	Iran (Persia)
Rhamnaceae	Ber (Chinese jujube)	*Ziziphus jujube*	Drupe	48	China
	Indian jujube	*Ziziphus mauritiana*	Drupe		
Rosaceae	Almond	*Prunus communis*	Drupe	16	Central Asia
	Apple	*Malus x domestica*	Pome	34	South Western Asia
	Apricot	*Prunus armeniaca*	Drupe	16	North Eastern China

	Loquat	*Eriobotrya japonica*	Pome	34	Central East China
	Quince	*Cydonia oblonga*	Pome	34	Caucasus region
	Peach	*Prunus persica*	Drupe	16	China
	Pear	*Pyrus communis*	Pome	34	Western China
	Plum	*Prunus domestica*	Drupe	32	China
	Sour cherry	*Prunus cerasus*	Drupe	16	South East Europe
	Sweet cherry	*Prunus avium*	Drupe	16	Asia minor
Rosaceae	Strawberry	*Fragaria x annanasa*	Etaerio of achenes	56 (8x)	France
Rutaceae	Bael	*Aegle marmelos*	Amphisarea	18	India
	Mandarin	*Citrus reticulata*	Hespe-ridium	18	South Eastern Asia
	Satsuma mandarin	*Citrus unshiv*	Hespe-ridium	18	South Asia
	Sweet orange	*Citrus sinensis*	Hespe-ridium	18	Indo- China
	Acid lime	*Citrus aurantifolia*		18	Iran (Persian)
	Grape fruit	*Citrus paradisi*	Hespe-ridium	18	West Indies
	Rough lemon	*Citrus jambheri*	Hespe-ridium	18	South East Asia
	Lemon	*Citrus limon*	Hespe-ridium	18	South East Asia
	Wood apple	*Feronia limonica*	Amph-isarica	18	India
Vitaceae	Grape	*Vitis vinifera*	Berry	38	Western Asia and Central Europe
Tiliaceae	Phalsa	*Grewia subenaequalis*	Drupe	36	India
Sapindaceae	Litchi	*Litchi chinensis*	Nut	30	South China
	Rumbutan	*Nephelium lappaceum*	Berry	22	Malayan Archipelago
Sapotaceae	Sapota	*Achras zapota, Manilkara hexendara*	Berry	26	South Mexico
Juglandaceae	Walnut	*Juglans regia*	Nut	32	Central Asia
	Peacanut	*Carya illinoensis*	Nut	32	North America
Leguminosae	Tamarind	*Tamarindus indica*	Pod		India

5

Nursery Management

Nursery: Nursery is a place where seedling, saplings or any other planting materials are raised, propagated, multiplied and sold out for planting.

Importance of a Nursery

1. The young seedling requires special attention during the first few weeks after germination. It is easier and economical to look after the young and tender seedling growing in nursery bed in a small area than in the large permanent site.
2. Majority of fruit crops are propagated by vegetative means. The propagules require special skill and aftercare before transferring them in the main field in a controlled condition in nursery, all these can be provided successfully by skilled labours.
3. Direct sowing method is not successful when compared with transplanting of seedling raised in nursery.
4. Nursery hardened plants are preferred for causality replacement in orchards.
5. Besides these, raising of seedling or sapling in nursery provides more time for replanting preparations.
6. Seasoning of seeding against natural odds is only possible in nursery.

Classification of Nursery

Nursery may broadly be grouped into two on the basis of its size:

1. Home Nursery
2. Commercial Nursery

1. **Home Nursery**: Home nursery is the area where planting materials are specifically grown only to eater the needs of the grower's garden. The area is small and the primary consideration is the raising of quality materials. Costly methods of nursery practices are adopted.
2. **Commercial Nursery**: The commercial nursery is mainly concerned with economic return from the investment and, therefore, very expensive nursery

practices are avoided without affecting the quality of the products. This type of nursery can be subdivided into two groups which are as follows:

a) **Rural Nursery**: Rural nursery is situated in a village near the high road or railway station. In general, the size of rural nursery may be large as the land and labour charges are cheaper. The products are also sold at a cheaper rate.

b) **Urban Nursery**: The nurseries which are located inside the city or close to it are known as urban nurseries. As the land is very costly and not easily available, the size of this type of nursery is usually small. The labour charges, transport cost etc. are also very high, but these are compensated by the higher price of products and volume of sale. Sometimes they act as middlemen i.e., procure planting materials from rural nursery and resale to the customers.

Factors Affecting the Establishment of a Nursery

1. Location and Site

a) **Topography**: The land for the nursery should be as flat as possible with a gradual slope in one direction. This helps in quick draining out the excess water during the heavy down pour. The area should be somewhat elevated than the adjoining area to avoid water logging and congestion.

b) **Climate**: It is of prime importance to select only those species and varieties which thrive well in the region.

c) **Reputation of locality for nursery business**: The locality where a nursery is intended to be established should have the reputation for the business. This means that the nurseries should be established in the known locality. Since such a place is known to the customers, the fledgling nursery gets the advantage of publicity. Apart from this, skilled labours and various inputs are also readily available.

d) **Transport facility**: Adequate transport facilities are of prime importance for successful nursery business. A nursery should be located near a metal road, or close to a railway station to deliver the plant materials to the customers in a short time. It should be easily accessible to the customers, but totally free from pilferage.

2. Selection of Soil

Type of soil, drainage and soil fertility are the three basic components to be considered while selecting the soil for establishing a nursery. Almost all the fruit crops do well in friable, loamy soil, rich in organic matter with the soil reaction varying from slightly acidic to near neutral (pH 5.5 to 6.5). However, a clay-loam soil is advantageous as in this type of soil, it is easy to lift the evergreen plants with earth balls intact. Shallow soil with hard and compact subsoil's should be avoided as it creates hindrance in proper root growth of mother-plants. A depth of 75.0 cm is sufficient for nursery practices but for the progeny block a soil depth of 1-1.5 m is desired. Presence of calcium carbonate layers affects permeability and aeration, and results stagnation of water which is detrimental to root respiration.

3. Water Supply

Nursery stocks require frequent light irrigation. A regular water supply must be assured in a nursery. The water for irrigation should be free from any dissolved salts and pH should be near neutral. A surface well may be dug or a tube-well may be installed for this purpose. In canal-irrigated areas, it is advisable to construct a water reservoir, so that the requirements may be met with the stored water during the dry period.

4. Manures

For nursery work, plenty of organic manures viz. farm yard manure, compost, leaf mould, sludge etc. are required. These should be available in sufficient quantities.

Parts of a Nursery

A nursery should consist of the following parts

1. **Building Structures**: The building structures include office, sale counter, packing shed, potting shed, store, implement shed, bullock shed and residential quarters etc.

Office: The office is the most important among all the building structures because all the business transactions are done from the office. The size of office depends on the size of business.

Sale counter: The sale counter should be attached to the office.

Packing shed: Before delivering the material to the customers they are to be packed and labeled properly. Packing materials like baskets or boxes are kept in this shed. It should be located near the sale counter.

Potting shed: Pot of varying sizes and shapes and the potting mixture (e.g., loamy soil, leaf mould, well rotten cowdung, fine river silt, sand) are kept in the shed. Potting operations are executed under shade where all the materials are easily available in right condition throughout the year. Only the plants are brought from the nursery beds. Potting under shade also minimises desiccation of plant, saves time and helps to keep other areas of the orchard clean.

Store : A store is needed to keep the fertilizers, sprayers, insecticides, fungicides etc. This should be located inside the nursery and away from the main office.

Implement shed : A shed is required for keeping the garden implements.

Bullock shed: Many bullock drawn implements are generally used in a small nursery. In such case, a shed is necessary and should be located in the rear side of the nursery to maintain the bullock herd of the nursery

Residential quarters : Sometimes the owner desires to live in the nursery premises with his family. In such case, the house should be constructed according to the choice of the owner. In large enterprises, residential arrangements are also made for the skilled labours and other staff members.

2. **Progeny Tree Block** : The correct choice of kind and variety of fruit crops and collection of true-to-type mother plants has strong bearing on the success and goodwill of a nursery industry. Suitable fruit crops should be selected to meet the demand of the customers. There should be a collection of good number of promising varieties of popular crops to make a wide choice. The progeny trees should be healthy, disease free, genetically true-to-type and free from insect-pests attack. The pedigree of these plants should also be known to the nursery man.

3. **Propagation Structures** : The outdoor conditions may not be always congenial for successful multiplication and raising of plants. The freezing cold atmosphere in winter and hot desiccating wind in summer preclude the outdoor cultivation of propagules. Different types of propagation structures green house (glass house), hotbed, cold frame, lath house, net house and mist chamber are in commercial use to create propitious conditions in respect of light, temperature and humidity for facilitating germination of seed, rooting of cutting and also for hardening of young seedling before transplanting them to main field.

Green house : A green house is a structure covered with glass for protection against adverse climatic conditions to cultivate plants. Plants are grown or cuttings are rooted in a green house when the atmospheric condition does not permit for raising them in outdoor condition. Modern green houses are equipped with regulatory mechanism for controlling temperature, light intensity, air flow and humidity. To minimise the high cost of glass, it may be replaced by polyethylene films of various thickness. Polyethylene of 0.10 to 0.15 mm (0.004 to 0.006 inch) thickness, resistant to ultraviolet ray is found most suitable for green house. For controlling the temperature and humidity inside the green house, arrangement should be made for proper air movement. Conservation of heat is essential in the green houses. In general, green houses are heated by steam or hot water passing through pipes fitted suitably from a central boiler. In large green houses, heated air is often blown into large (30 to 60 cm) polyethylene tubes having small (5 to 7.5 cm) holes spaced throughout the length to release the hot air inside, which are hung overhead along the length of green house. The heat loss in green house can be reduced by providing a double layered, sealed polyethylene sheet outside the glass.

Plants, in green house, are grown in the ground bed or bench type raised beds. The size of the green houses and the dimensions of beds vary according to the needs of the propagules.

Hotbed : It is a small fixed structure having three components namely, frame, cover and heating unit. It is also used for growing small tender seedling and rooting of cuttings. The cover is made of glass or transparent polyethene sheet. The soil of bed is dug out and replaced with 30-50 cm thick layer of raw cowdung, preferably horse manure and covered with 15-20 cm thick rooting medium. Decomposition of raw dung generates heat and the temperature of the rooting medium is raised. In large size modern nursery, improved heating arrangements like steam piping, hot water piping, electric cable etc. are used for heating the bed. The frame can be made of wood or any light metal.

Cold-frame : It is a small and movable structure of glass which encloses a ground bed. It is used for hardening or conditioning of rooted cutting and young seedlings before transferring them out. There is no bottom beating arrangement and heat is tapped from the solar energy. It should, therefore, be located in a sunny place. For assured success, care has to be taken in respect of proper ventilation, shade, watering and winter protection. The sash is removed during day time for quick raising of temperature of growing medium

Lath house : Lath house is a structure which is made with the object of providing shade and thereby protecting the young tender seedlings or rooted cuttings which are sensitive to strong sunlight and high temperature. The transpiration and evaporation loss of moisture are directly related with temperature. So, in lath house the loss of moisture from leaf as well as from soil surface is greatly reduced, due to low temperature and light intensity, resulting into less water requirements of plants. Seedlings are transferred for hardening from green house or hotbed to lath house. The structure and size of lath house vary widely. Different kinds of shading materials are used (e.g. wood strips, polypropylene fabrics etc.) for the roof.

Net house : A net house resembles to a green house where the roof as well as sides are made of wire nets fixed on iron or wooden frame. The roof may be covered by polyethylene sheet, leaves or some climbers may be allowed to grow and cover the roof to provide shade. A net house is used for growing shade loving plants, orchids, cactus and also for raising seedlings. The inside atmosphere remains cool in summer and warm during winter.

4. **Nursery Beds** : A nursery bed should be prepared by repeated ploughing and pulverizing the soils to obtain a fine fifth. Adequate amount of organic manure and compost should be incorporated into the soils. Nitrogen at the rate of 25 kg per hectare may be added for quick growth of seedling at the initial stage. Addition of wood ash makes the soil loose and acts as a source of potassium which is required for better growth. A loose, friable, clay loam soil, rich in organic matter is ideal for the young tender seedling. Proper drainage system must be assured because the young seedlings are sensitive to water logging condition. Nursery beds may be little raised during rainy.

Management of Young Nursery Plants

a) **Irrigation** : The young tender seedlings are exacting in their water requirement. Over watering is as harmful as under watering. To maintain the turgor pressure in the cell for maximum photosynthetic activity, light irrigation should be given at an interval of 7 to 10 days during summer and 15 days in winter. Proper care has to be taken to avoid subsoil congestion.

b) **Nutrition** : Proper nutrition has profound influence on the growth of nursery plants. Liberal manuring with well decomposed organic matter is a prerequisite for achieving success in nursery business. A light and frequent does of nitrogenous fertilizer @25 kg per hectare may be applied to obtain quick growth of seedling.

c) **Weed Control** : Weeds poses severe threat to the young seedlings in the nursery bed. Weeds being wild in nature grow vigorously and get established earlier than the crop plant. Often the seedlings are crowded over by the weeds if they are not controlled initially. The nursery bed should be kept free from weeds by shallow hoeing. This also helps in better aeration of the soils. The roads and irrigation channels also should be made free from weeds to avoid the chances of spread of weed seeds.

d) **Plant Protection**

   i) **Protection against adverse climatic condition:** During the daytime in summer season, temperature shoots up in tropical regions resulting into striking rise of transpiration and evaporation rates, If high temperature in accompanied by flew of hot wind, the problem is further aggravated and the tender seedling soon losses its vitality. The tolerance limit of fruit plants to high temperature varies with species.

Seedlings can be protected from scorching sun by providing some shades during day time. Thatches may be erected over the seed bed. Shade crops may also be grown. Windbreaks on the eastern and southern sides of the nursery can reduce the wind velocity, and thereby, minimise the moisture losses from leaf and soil surfaces as well. Recuperation of soils with frequent light irrigation during summer months greatly helps to overcome this problem.

ii) **Protection against pests and diseases:** The reputation of a nursery depends on the supply of genuine, healthy and disease free plants. Once the seedlings are affected by pests and/or diseases, their growth is arrested and they turn pale and weak. The tender seedlings are susceptible to various pests and diseases but the degree of susceptibility varies with species and age of seedlings.

e) **Lifting and packing of seedling** : It is desirable to lift the plants with balls of earth intact while the deciduous plants can be lifted bare rooted during dormant condition. Light irrigation should be given, if required, before lifting the seedlings. The type of packing depends on the size of stock and type of plant. The transplants lifted bare rooted are packed with their roots wrapped in moist sphagnum moss and moist paper. The ball of earth may be wrapped with coconut leaves or rice straw and packed in basket.

f) **Storage** : Low temperature (0.5 to 1.5^{0}C) in association with high humidity (80-90%) in the atmosphere of storage reduces the respiration and transpiration rates. The store house should be well ventilated to drive out the carbon dioxide and heat of respiration.

6

Plant Propagation

Plant propagation: Plant propagation means multiplication of plant by sexual or asexual method.

There are two type of plant propagation

1. Sexual propagation
2. Asexual propagation

A. Sexual propagation: Sexual propagation means multiplication of the plant by seed. It is also known as seed propagation.

Advantages of sexual propagation

1. This is the easier and cheaper method.
2. Some of the plant does not response well by asexual method (Papaya) this type of plant only propagated by sexual propagation.
3. Sexually raised plant are long lived have extensive root system and bear heavily as compare to asexually propagated plants.
4. Hybrid can develop only by seed propagation.
5. Seed can be store for longer duration.

Disadvantages of sexual propagation

1. Most of the fruits plants are heterozygous in nature.
2. Seedling raised plants are generally tall and spreading type causing hazard for carrying out various management practices.
3. Sexually raised plant take longer duration to come bearing.
4. Seeds of the many fruits are to be shown immediately after extraction, because they loss their viability soon eg. Phalsa, Jackfruit and Mango etc.

B. Asexual propagation: Asexual propagation means propagation of plant by vegetative part like stem, root, leaf etc.

Advantages of asexual propagation

1. Most of the fruit crops are heterozygous in nature. Genetic architecture can only be maintained true to type by asexual mean.
2. Some of the fruit plants do not produce viable seed (Banana, Pineapple), multiplication of them is only possible through asexual means.
3. Vegetativelly propagated plants are early bearing.
4. Height of plant can be regulated and resistance of biotic and abiotic stress can be developed by use of suitable rootstock.
5. Bridge grafting-repairing of the plants.

Disadvantages of asexual propagation

1. Vegetative propagation does not evolve new varieties.
2. Some time it is expensive and it require special technique.
3. Vegetativelly raised plants are short lived.

Methods of asexual propagation

A. Apomictic Seed- Citrus, Mango etc.

B. Specialized vegetative structure- Bulb, Corm, Rhizome, Tubers, Suckers, Runner and Crown

C. On the own root system-

i. **Cutting-** Herbaceous, Soft wood, Semi hard wood and Hard wood.

ii. **Layering-** Simple, Tip, Trench, Air layering, Mound layering and Serpentine layering

D. On the root system of other plant-

i. **Budding** – T-budding, Ring budding, Forcurt budding, Chip budding, Patch budding, Flute budding, I- budding etc.

ii. **Grafting**

a. Attached method of grafting – Inarching

b. Detached method of grafting- Veneer grafting, Toung grafting, Splice grafting, Whip grafting, Epicotyle grafting etc.

i. Cutting

Cutting is piece of plant which produce adventitious root (in case of stem and leaf cutting) and a new shoot system (in case of root cutting) when placed in rooting media keeping all other condition are favourable. It gives rise to a new plant almost true to mother plant.

Advantages of cutting

1. It is very easy to perform and economical.
2. Rapid multiplication is possible within a short time.
3. Does not require much space.
4. Complicated stionic relationship can be avoided.
5. Cutting can be transported to any place because the stem or root piece remain fresh for 5-7 days if packed properly.

Disadvantages of cutting

1. All the plant does not root very easily.
2. The benefit of root-stock cannot be exploited.

Classification of cutting

1. **Stem cutting-** Herbaceous cutting, soft wood stem cutting, semi hard wood stem cutting and hard wood stem cutting
2. **Leaf cutting**. Not common in fruit crops.
3. **Leaf bud cutting**- Not common in fruit crops.
4. **Root cutting-** It is more common in forestry crop

Length of cutting- 8-15 cm with 2-3 nodes and diameter of cutting-1-2 cm

Time of cutting

i. Deciduous plant- March-April
ii. Evergreen plant- May-September

Climate of cutting- Average temperature-20-25^{0}C, day temperature-21-27^{0}C, night temperature-15-16^{0}C and humidity-80% and rooting period-15-30days.

Types of stem cutting

1. **Herbacious cutting**- Herbaceous cutting are tender succulent leafy part of stem of herbaceous plant. The terminal 7-12 cm of shoot is cut and basal leaves are removed and upper leaves are undisturbed.
2. **Soft wood cutting**- Soft wood cutting are those prepare from soft succulent terminal portion of woody perennial. It is desirable to retain some leaves for photosynthesis but more leaves are undesirable which may cause moisture loss due to transpiration. Soft wood cutting are 8-15 cm long with 2-3 nodes. In deciduous plant cutting should be taken in March-April. In evergreen plant cutting should be taken in May-Sept. Rooting in cutting within 15-30 days if temperature 20-25^{0}C & 80% humidity.

3. **Semi hard wood cutting**- Prepare from partially mature but tender woody shoot. Wood cutting is taken from lignified nature woody shoot. Length may vary from 20-25
4. **Hard wood cutting-** Hard wood cutting having diameter from 1 to 2 cm. each cutting has 2-3 nodes. All the cutting performs well in misting of water and exogenous application of auxin.

Factor affecting the rooting of cutting

1. **Water**-When cutting is detached from mother plant it natural supply is stopped but transpiration is continued. Resulting severe water loss from lower surface. For rapid and better regeneration of root it is necessary to maintain turgidity of cells. Transpiration can be minimized by high humidity around the cutting.
2. **Temperature-** Day temp-21-27^{0}C and night tem-15-16^{0}C. Some difficult to root plant with bottom heat technique keep the rooting medium worm.
3. **Ligh**t- Long day has been found to for the regeneration of root due to more accumulation of carbohydrates.

   Light condition-Shooting is more

   Dark condition-Rooting is more
4. **Rooting medium**-Ideal rooting media loose, porus with high water holding capacity. pH should be neutral.
5. **Hormone**-Among the various growth hormone auxin have profound influence on root formation. In general plants have the two hormones i.e. IAA and IBA which are responsible for rooting in cutting. IAA is always in natural form whereas the IBA which is natural form and its synthetic form like Rootex, Linolooll pest are available in the market. These rooting hormone IAA and IBA always form the tip of the plant and move toward the basal portion. The movement of auxin from upward to downward is known as polar movement. In case of cutting for proper rooting we should always maintain the polarity of the cutting. For the identification of the upward and downward of the cutting the upper portion should be slant cut whereas, the basal portion should be strait.

ii. Layering

Layering involves the regeneration of adventitious root from a shoot when it is still attach with the mother plant. The exo-genus application of auxin has been found to hasten the initiation of root, increase percentage of rooting and number of roots per layers. This is more common in guava, litchi, citrus etc.

Advantages

1. It is easy method does not require much care and arrangement like cutting.
2. Mother plant supplies nutrient and it remain attached while rooting.
3. Some of plant species which do not root by cutting can be propagated by layering.

Disadvantages

1. The number of new plant can produce from any given mother plant by layering is low as compare with cutting.
2. Costly method where labour charged is very high.
3. Layered plants are generally shallow rooted.
4. Benefit of root stock cannot be exploited.

Type of layering

1. Simple layering.
2. Tip layering.
3. Trench layering.
4. Air layering
5. Mound or Stool layering.
6. Compound layering.

iii. Grafting

Grafting is the art of uniting or joining the parts of two independent plants in such a manner that they unite together and develop into a single independent plant. The part of graft which is to become the shoot system is termed as scion. The part which is to become the root system is called as root stock.

Types of grafting

- Approach grafting
- Inarching
- Bark grafting
- Cleft grafting
- Splice grafting
- Root grafting
- Bridge grafting
- Side grafting
- Veneer grafting
- Epicotyl grafting (Stone grafting)
- Soft wood grafting

iv. Budding

Budding is also method of grafting where only one bud with a piece of bark, and with or without wood, is used as scion material. It is also called as bud grafting. The plant successful union of the stock and bud is also known as budding.

Methods of budding: T-budding (shield budding), inverted T-budding, patch budding, flute budding, ring budding, chip budding.

Advantages

1. Root stock can be successfully utilized.
2. An inferior variety can be changed into superior type by top working.
3. Growing number of variety in one plant is possible.
4. Difficult to root can be successfully multiplied.
5. Early flowering and fruiting can be obtained.
6. The damaged plant part can be repair by grafting.

Disadvantages

1. All plants are not compatible to each other.
2. The method are confined only indicates plant having continue cambium, monocot and dicot have vascular cambium.

Table 6 : Different between cutting and layering

Cutting	Layering
In cutting the cut portion are remove from the mother plant initiation of roots take place when all condition are favourable.	In layering the initiation of roots in the plant parts when it is still attached with the mother plant when all condition are favourable

Table 7 : Different between budding and grafting

Budding	Grafting
In budding single bud is taken as a scion shoot	In grafting the length of stick is taken as a scion shoot which have more than one bud

Table 8 : Different between attach method of grafting and detach method of grafting

Attach method of grafting	Detach method of grafting
In attach method of grafting rootstock and scion shoot should be attached and after getting union scion shoot are detached with the mother plant	In detach method of grafting scion shoot are detached from the mother plant then union of rootstock and scion shoot occur

Propagation through Specialized Vegetative Structures

Various modified plant organs like bulb, corm, tuber, rhizome etc. which are primarily storage organ for food are used as planting material. Aerial part of these plant has die at the end of season while the underground part remain in dormant condition and resume their activity and sprout when prevailing climatic condition are favourable. There are certain plant modifications which are used for vegetative propagation of plants. These modified plant parts may be stem, root, or leaves and are usually specialized for food storage. Two principal methods are used for propagation of plants by using these modifications.

a. **Separation** : Naturally detachable structures, such as bulbs or corms are separated and planted individually e.g. bulb, corm

b. **Division**: The plants modification such as rhizomes, tubers etc., are cut into sections to obtain new plants from each section eg. rhizome, tuber.

Bulbs: A bulb is a storage organ consistory of a short flattened stem bearing growing point at the apex enclosed fleshy scale leaves. When the bulb place in optimum condition adventitious roots are emerging out. Bulbs are produced by monocotyledonous plants in which the stem is modified for storage and reproduction. Bulb is a specialized underground organ consisting of a short freshly, usually vertical stem axis bearing at tip apex or growing points and enclosed by thick freshly scales. Bulb scales morphologically are the continuous sheathing leaf base. Growing points develop in the axils of these scales to produce miniature bulbs known as bulblets/daughter bulbs. These daughter bulbs can be separated from the mother plant at the end of growing season and used as propagating material eg. Tulip, Daffodils, Tuberose, Onion, Garlic, (cloves).

Tubers: Tubers are swollen underground modified stem containing food reserve and function an organ for vegetative propagation. Tuber consist all part of stem i.e. node, internodes, lateral and terminal buds. The eyes of tuber are arranged spirally representing nodes. A tuber is the short terminal portion of an underground stem which has become thickened because of accumulation preserved food material eg. Potato. Propagation by tuber can be carried out either by planting the whole tuber or by cutting into sections each containing bud or eyes.

Tuberous roots: Fleshy modified root active an storage organ bearing growing point on the crown or stem end without node and internodes. Certain herbaceous perennials produce thickened roots which contain large amount of stored food. The tuberous roots differ from the tubers in that they lack nodes and internodes. Adventitious buds are present only at stem end or proximal end; fibrous roots are produced towards the distal end. These fleshy roots are separated and used for propagation eg. Sweet potato, Dahlia, Tapioca (Cassava).

Rhizomes: Botanically rhizome is an elongated horizontal fleshy underground stem having node and internodes bearing bud in the axial of reduced scale leaves. Rhizome are cut into piece each with one or two viable bud and plated in favourable atmospheric condition. The horizontal, thick and fleshy or slender and elongated stem growing underground are known as rhizomes. Rhizomes have nodes and internodes and readily produce adventitious roots. The rhizomes are cut into pieces, each containing vegetative bud and transplanted eg: Banana, Ginger, Ferns, Turmeric, and Cardamom.

Corms: Corm is short solid swollen underground base of stem axis having distant node & internodes and enclosed by dry scale like leaves. A corm is solid underground base of a stem having nodes and internodes and is enclosed by a dry scale like leaves. After flowering one or more corms may develop just above the old one, which disintegrates. In addition several new corms called caramels develop below each new corm. These may be separated and grown for 1-2 years to reach flowering stage eg. Gladiolus, Alocacia.

Cormels- Cormels are small corm develop in between old and new corm which require two year growth to come into flowering stage

Runners: A long prostrate shoot arise from the leaf axils and give rise to a new plant at the end. The runner production is formed by long day & high temperature. Runners are specialized areal stems (stolones) arising in the leaf axils of plant having rosette crowns. New plants arise from nodes at interval along these runners. From these runners more new runners may arise thus developing natural clonal multiplication methods. The typical runner producing plant is straw berry which is photo sensitive with regard to its runner production. Long days favour runner production where as short days prevent runner formation eg. Strawberry.

Suckers: A shoot arising as an old stem or underground part of stem or on horizontal root system called sucker. Adventitious shoot from the underground portion of the stem or from their horizontal root systems are known as suckers and when these strike roots, they may be utilized as propagation materials. Well developed suckers are dugout and separated from the mother plant and planted in the nursery for further growth. Suckers are usually treated like rooted layers eg. Pineapple, Banana.

Crown- A crown is a very short condensed part of plant at the surface of ground which gives rise to new shoots, crown grown in top of the pineapple fruit. It is a vegetative growth attached to the central core of fruit which is use as planting material.

Offsets/offshoots: An offset is a shoot or thick stem of rosette like appearance arising from the base of the main stem of certain plant such as date palm, pineapple etc.

- Date palm cultivars are propagated vegetative by separating away the offshoots and replanting them.
- However these are girdled and layered for about a year prior to separation, because offshoots do not root easily when directly separated from the mother plant and planted in the field.

Micro propagation- Multiplication of plant in a septic condition and in artificial growth medium from very small plant part like meristem tip, callus, embryo, another etc. Haberlandt German plant physiologist in 1902 first describes the biological principle of tissue culture.

Merits

1. Tissue culture help in rapid multiplication of true to type plants throughout the year.
2. A new plant can be regulated from a miniature plant part where as in conventional method a shoot of considerable length is required.
3. Large number of plant can be produce in culture to be in small space with uniform growth and productivity instead of growing them in large area in nursery.
4. Plant raised by tissue culture is free from disease.
5. Tissue culture coupled with somatic hybridization helping evolving new cultivar in a short time.
6. Micro-propagation facilitates long distance transport of propagation material and long storage of clonal material.
7. Tissue culture method is path clearly effective in plants that do not weed free from seed. Seeds are not viable (male sterile) or available easily.

Demerits

1. Cost involves in setting and maintenance of laboratory is very high and may not insanity their use is all the horticultural plant ordinarily.
2. Tissue culture technique requires small manpower.
3. Slight infection may damage the entire lot of plants.
4. Seedling grown under artificial condition may not survive when placed under environmental condition.

7

Orchard Planning and Management

Orchard: Establishment of orchard is a long term investment. It is require a proper planning, any mistake during the selection of site, planting distance, choice of crop/variety, quality of nursery stock etc. reflect greatly on the orchard performance and efficiency.

Following points are keep in mind at time of orchard planning-

1. **Road-** Orchard should be well connected with the road for facilitate transportation of material to the orchard and produce of orchard to the market.
2. **Orchard structure-** In structure office, implementshed, go-down come store etc. are required. Area under these structures is not more than 10% of total area.
3. **Fencing and wind break-** Before planting of plants, installation of fence is necessary for protecting the undesirable element. Fencing can be done by erecting a wall or by fixing iron angle or concrete cement pillar and titening with running 4 to 5 times at a distance of 30-40 cm barbed wire.

   Fencing can be done by two ways:

   1. By barbed wire.
   2. By suitable hedge plant.

   Height of the hedge can be maintained with regular pruning. Pruning should be done twice in a year. The suitable plants are- *Carissa carandus, Lawsonia alba, Chloredendran eneremis, Durenta plumeri, Parkinsonia aculeata, Acacia spp., Zyziphus spp., Pithecellobium dulsis* etc.

   Wind break- Orchard tree should be protected from high velocity wind which could harm by uprooting tree, breaking branches, causing pre-maturing fruit drop, erosion of top soil and evaporation of soil moisture.

   Suitable plants are- *Casurina equisitifolia, Gravillea robusta, Polianthia longifolia, Acacia aculiformis, Carissa carandus, Putranjiva roxburghai, Dalbergia sisso* etc.

4. **Irrigtion:** For irrigation source of water should be at the centre of the orchard.
5. **Demarcation of block-** Block should be demarcated according to nature of the plants. The water requirement of the plant, cultural requirement of the plants, training pruning requirement of the plant etc. are different i.e.why the specific nature of the plant should be demarcated in seperate block.
6. **Spacing-** Proper spacing should be keep provide favourable condition to root and shoot development. An optimum distance between row and within the row is most essential. Closely planted tree grows poorly and give poor yield and quality of fruits, suffer more with disease and pest. Planting distance varies with climatic condition, soil type, growth habit, root stock use etc. in general there are two type of growth habit i.e. spreading type and erect type. The spreading type plants required more spacing whereas, the erect type plant required less spacing. Some of the important fruit crops and there spacing are as follows:

Crop Plants	Distance
Mango, Jamun, Litchi, Ber	10×10 or 12×12
Jack fruit and Bread fruit	12×12
Date palm, Fig, Mandarin, Lime, Lemon	6×6
Pummelo, Grape fruit	6 ×6 or 7×7
Guava and Cashew nut	6 ×6 or 8× 8
Sapota, Loquat, Avacodo	8× 8 or 9× 9
Custard apple	4.5 ×4.5

7. **Selection of planting material-** At the time of selection of planting material the following points are keep in mind.

- They should be true to type
- They should not have been budded or grafted not more than 20-30 yrs. old plant.
- Planting material should have to certified by competent authority.

8. **Planting of fruit tree-** At the time of planting of fruit tree following points are keep in mind-i.e. growth behaviour, canopy development and fruiting habits. The fast growing tree should be planted in separate block whereas, the slow growing tree planted separately. In canopy development the erect type of plant and spreading type of plant are planted in separately in separate block. Regarding fruiting habit there are two types of plant i.e. terminal bearing and axillary bearing each type of plants are planted separately.

Orchard Management

- Orchard is an area, often enclosed, devoted to the cultivation of fruit trees and as a unit it encompasses various resources like land, water, trees and external inputs.

- All these resources have to be well utilized to the best advantage for higher production per unit area on sustainable basis without adversely affecting the quality of environment.
- We should also understand that a good manager is one who gets maximum output of various inputs consistently without any loss of fertilizers and manure, plant, plant protection chemicals, produce etc.
- Therefore, one should understand the management of these qualities of both resource and output.

Resources for better comprehension of orchard management are;

1. Soil management
2. Water management
3. Nutrition management
4. Pruning and training (plant management)
5. Weed management
6. Plant protection against insect pests and diseases.
7. Bearing, fruitfulness and causes of unfruitfulness.
8. Maturity and harvest.
9. Post harvest handling, utilization and marketing.

Soil Management

- Soil management aims at maintaining soil in good condition, or improving the condition if necessary.
- This includes protection from direct sunlight and from the impact of rainfall and wind erosion.
- In annual crops like vegetables and flowers which do not leave vacant space.
- There is no such problem except that one has to replenish nutrients harvested by crops and leached out but in tree crops, wherein, it is usually several years after planting before a tree which form such an extensive canopy that it can provide adequate protection to the soil, the vacant space needs to be productively utilized and protected through different management practices like intercropping, cover cropping, cultivation, sod culture, mulching, rotation, high density planting.

Objectives of soil management

1. To create favourable conditions for moisture supply and proper drainage.
2. To maintain high fertility level and replenishment against losses.

3. To provide proper soil conditions for gaseous exchange and microbial activities through addition of organic matter.
4. To check or reduce soil erosion.
5. To ensure supply of nutrients for growth and development of plants.
6. To utilize vacant land for additional income because such a loss is inconceivable for small holders.
7. To reduce the cost of cultivation with high economic returns.
8. To suppress weed population.

Definitions of terms to be used in management of soil

1. **Intercrop:** Any crop other than main crop grown between the rows of perennial tree crops is known as intercrop and the cultivation there of is intercropping.
2. **Green manure crop:** The crop other than main crop grown for the purpose of enriching the soil for organic matter is called green manure crop.
3. **Cover crop:** The crop grown to provide a cover to soil to protect it from erosion. It may be green manure crop also.

Methods of soil management: Appropriate soil management method is important for the control of weeds, incorporation of organic and inorganic fertilizers and to facilitate absorption of water in soil.

The common soil management practices are,

1. Cultivation
2. Sod Culture
3. Mulching and
4. Rotation

Choice of the system is determined by many factors as mentioned below:

1. Type of crop
2. Rooting depth of the crop
3. Slope of the soil
4. Rainfall of the area
5. Climatic condition of the place
6. Economic condition of the farmer

8

Planting System

Layout plan- Main objective of layout is maximum number of plant accommodated in per unit area without affecting the efficiency of production. If more than one fruit is included in the orchard plan, each kind should be grouped into individual block. Similarly fruit ripened in particular season should be grouped to facilitate convenience in watch and ward, harvest and post harvest handling. Some kind of fruit trees need pollinators/pollinisers for improving the fruit set.

1. Square system
2. Rectangular system
3. Triangular system
4. Hexagonal system
5. Quincunx system
6. Contour system

The following are the important systems of planting generally followed on the basis of agro-climatic conditions to improve aesthetic view of the land.

1. Square System

It is the most commonly used method and easy to layout in the field. In this system, plant to plant and row to row distance is the same. The plants are at the right angle to each other, every unit of four plants forming a square. This system facilitates the interculture in two directions after the orchard is planted.

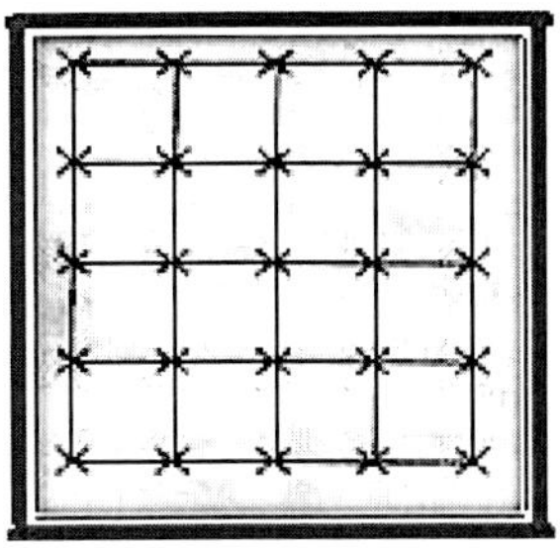

Advantages

1. Most easy and popular one.
2. In this row to row and plant to plant distance is kept similar.
3. Plants are exactly at right angle to each others.
4. Intercultural operations can be done in both the directions.
5. Adequate space for inter-cultivation of remunerative crops like vegetables.

2. Rectangular System

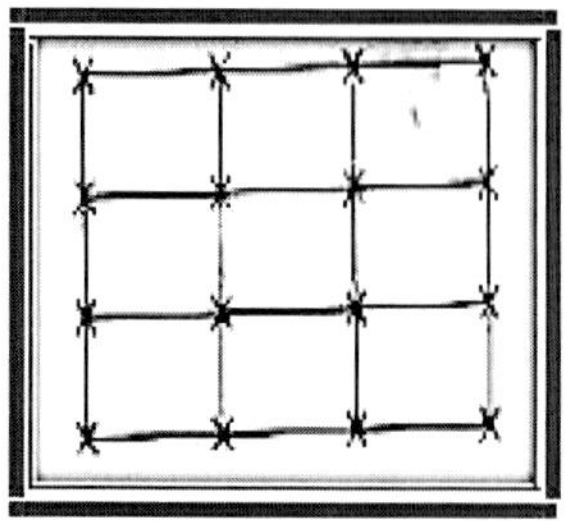

In this system, the plot is divided into rectangles instead of squares and trees are planted at the four corners of the rectangle in straight rows running at right angles. Like square system, this system also facilitates the interculture in two directions. The only difference is that in this system more plants can be accommodated in the row keeping more space between the rows.

Advantages

1. Lay out in rectangular shape.
2. More space between row to row.
3. Inter-cultural operations can be done in both the ways.
4. Plants get proper space and sunlight.

3. Hexagonal System

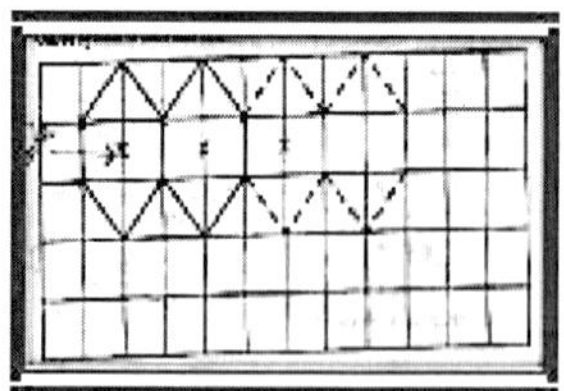

In hexagonal system, the trees are planted in the corners of equilateral triangles. Six trees thus from a hexagon with another tree at its centre. This system, through a little difficult for execution but accommodates 15 percent more plants than square system. Cultivation of land between the tree rows is possible in three directions with this system. This system is generally followed where the land is costly and very fertile with ample provision of irrigation water.

Advantages

1. Accommodates 15 % more plants than the square system.
2. Plants are planted at the corner of equilateral triangle.
3. Six trees are planted making a hexagon.
4. The seventh tree is planted in the centre and called septule.
5. This requires fertile land.

Disadvantages- It is the complex nature of layout and more number of plants of plant population causing more competition for moisture and nutrient.

4. Quincunx System

This is also known as diagonal system. It is similar to square system except that fifth plants located at the centre of each square term as filler plants as a result

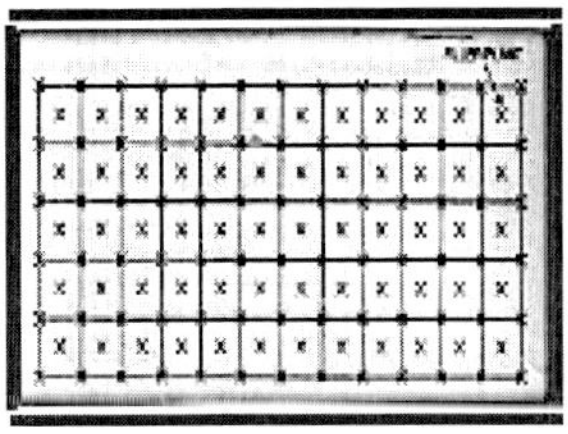

number of plant per unit area are become about an and half time more than square system. The number of plants per acre by this system is almost doubled than the square system. Fruit trees like papaya, kinnow, phalsa, guava, peach, plum etc. can be planted as fillers in the permanent trees provides an additional income to the grower in the early life of the orchard. The filler trees are uprooted when the main orchard trees start commercial fruiting.

Characteristics of filler plant

1. Short duration
2. Quick growing
3. Erect type
4. Early maturing-eg.-Banana, Papaya, Pomegranate.

5. Contour System

This system is usually followed in the hilly areas with high slopes but it is very much similar to the square/rectangular system. Under such circumstances, the trees may be well planted in lines following the contour of the soil with only a slight slope. Irrigation and cultivation are then practiced only across the slope of the land as this practice reduces the chances of soil erosion. In this system layout is done as in square/rectangular system, first by establishing the base line at the lowest level and then marking for the trees should be done from the base to the top. Bench terraces are used where the slope is greater than 10 per cent.

6. Triangular System

1. In this system, trees are planted as in the square system but the plants in the 2nd, 4th, 6th and such other alternate rows are planted midway between the 1st, 3rd, 5th and such other alternative rows. This system provides
2. Plants in alternate rows are offset half the space between plants in a row.
3. Result in 9% fewer plants than square and rectangular system eg. Amrapali – 1600 plants / ha.
4. More open space for trees and for intercrop

9

Soil and Climate

Soil: Soil is the upper most crust of earth surface which supports plant growth. It is defined as a three phase system in which plants grow. These phases are solid, liquid and gas and are essential. Solid part is frame which provides space for other two. This consists of minerals, clay minerals and organic matter. The soil is also a living system with millions of microbes that breakdown organic matter and builds it again.

Some points are following

- Microbes are essential and survive only when soil is well aerated and rich in organic matter and avoid of waterlogged conditions.
- Texture of soil depends on the size of solid particles and classified as gravel, coarse and fine sand, silt and clay.
- Soils are classified according to relative distribution of these particles and there are 12 textural classes.
- Likewise, arrangement of these particles is referred as structure, and both texture and structure land soil physical properties like water holding capacity, aeration and bulk density.
- Generally loamy soils and crumb structure are most preferred for fruit crops.
- According to level of organic matter, soils are classified as mineral soil or organic soil and soil having more than 20% organic matter is organic soil like peat and muck.
- Minerals and salts land chemical properties to the soil like pH, alkalinity, solidity, salinity and cation exchange capacity which influence the availability of nutrients in soil.
- Therefore, for making choice for soil, soil analysis in terms of following criteria is essential to decide on land capability.

Criteria for land capability class

i. Slope and erosion hazard.

ii. Soil depth.

iii. Drainage.
iv. Workability.
v. Stoniness and rockiness.
vi. Water holding capacity.
vii. Permeability.
viii. Nutrient availability.
ix. Fertility status.
x. Salinity, alkalinity and acidity hazards.

- Based on these criteria there are 8 capability classes, of which (i) to (iv) are suitable for cultivation and (v) to (viii) are not suitable for cultivation.
- The soil provides support for the plant and act as storehouse of nutrients and water as well as oxygen for root growth.
- The ability of the soil to support plant growth is often referred to as its productive capacity which depends on fertility and physical condition. Therefore, the soil has to be a good soil.
- A good soil is one which has the capacity to nourish and sustain plant growth by providing mineral particles (nutrients) in an available form to plants by their interaction with soil air, moisture, microbes and humus.
- Generally a loam soil is considered to be a good soil.
- Generally fruit crops need porous, aerated, deep (2 m) uniformly textured soils and the pH of soil should be within range of 6-8.
- Soil with hardpan within 120 cm from surface, soil with high clay content at surface and very less at subsurface or vice-versa are not suitable for fruit crops.
- Fruit crops are susceptible to waterlogged condition and growth is adversely affected by salinity, solidity and alkalinity.
- It is, therefore, important that soil be analyzed for its quality and then choice of the crop is made for sustainable production.
- If the soils are problematic like poor aeration or drainage, solidity, alkalinity, acidity and salinity, they require improvement or reclamation before taking up crop production or the venture would fail.
- Alternatively tolerant or resistant crops can be chosen for different problems.

Salinity tolerant crops: Kair, Khirni, Woodapple, Date palm, Ber, Aonla, Fig, Sapota etc.

Sodicity tolerant crops: Ber, Tamarind, Woodapple, Date palm, Aonla, Karonda, Fig, Phalsa, Pomegranate, Guava, Bael and almond.

Drought tolerant crops: Ber, Aonla, Phalsa, Lasoda, Kair, Custard apple, Karonda, Fig, Guava etc.

- If we know the soil and the requirement of soil for the crops, then choice of the crop can easily be made.

Grouping of fruits according to their tolerance to salinity

a. **High salt tolerance:** Date palm, Ber and Aonla.

b. **Medium salt tolerance:** Pomegranate, Fig and Grape.

c. **Low salt tolerance:** Apple, Orange, Almond, Lemon and Avocado.

- In making choice of soil for fruit crops physical properties should be emphasized, more as chemicals can be added from outside to improve nutrient status and chemical properties of the soil.
- Generally the depth and the drainage-ability are very important for crop production.
- To upkeep soils for sustainable production following things are to be done before and after planting a crop:

Soil Analysis in Terms of Its Physical and Chemical Attributes

- Bring the soil to its optimum potential by applying organic matter, chemical fertilizers, micronutrient and amendments depending on soil analysis report.
- Adoption of soil conservation technique like green manuring on regular basis.
- Use of improved water management techniques like drip irrigation and check basin or Furrows.
- Incorporation of large quantity of bulky organic matter each year.
- Creation of appropriate drainage around the plot.
- Scrapping of salts and reclamation of soil by application of gypsum, iron pyrites, press mud etc., on regular basis in case of salinity problem.
- Replenishment of nutrients harvested by the crop on regular basis by preparing a balance sheet for nutrients.
- Recycling of organic waste.
- Soil is the most important natural resource for fruit culture and it needs to be protected and improved.

Climate: Climate is the most important factor on which choice of the crop for a region depends and therefore, understanding about soil and climate and their requirement for different crops for optimum production on sustainable basis is important for horticulturists.

Climate is defined as the whole of average atmospheric phenomena for a certain region calculated for a period of thirty years. These phenomena are light, heat, water and air.

Light

- Electromagnetic radiation to which the organs of plant react ranging in wavelength from 4000 to 7700 angstrom units, and is propagated at a speed of about 540 kilometres per second.
- It is essential for the process of photosynthesis and therefore, for growth and development of plants.
- There are two aspects of light, its intensity and duration which are important for plant development.
- The light intensity can be estimated from the number of hours of bright sunlight or from the cloudiness of sky.
- Generally horticultural crops need a lot of light and must be grown in sunny climate, but there are some crops which can tolerate shade eg. Turmeric and ginger.
- A third group requires permanent shade like salak palm, duku, and carambola.
- The duration of light for the time elapsing between dawn and dusk referred as photoperiod or day length. This exerts considerable influence on flowering.

Based on the response by plants the major classes are following. However, fruit crops for such categories are not known.

1. **Long day plants:** Passion fruit, Apple etc.
2. **Short day plants:** Strawberry, Pineapple, Apple, Pear, Peach, Plum etc.
3. **Day neutral plants:** Papaya, Guava, Banana etc.

Heat

- Heat is a non-mechanical energy transfer with reference to a temperature difference between a system and its environmental surrounding.
- It is measured as temperature by thermometers.
- The growth of the plants depends primarily on temperature.
- Availability of heat units decide the crop for a given place and the average temperature of a place gives an idea about heat units available on the basis of which crop can be decided.
- Temperate fruit crops like apple, pear, peach, plum and almond become dormant due to short day conditions in the region and need chilling of various lengths to break dormancy.

- Frost and chilling are harmful for tropical and subtropical plants.
- On the other hand extremely high temperatures found in arid region cause wilting, sunscald, necrotic spot and even death of plants.

Therefore, under such conditions appropriate choice of plants and provision of protection become important.

Based on temperature variations on the surface of the earth we have the following climates.

1. **Tropical:** Tropical equable climate with no distinct winter, eg. Mango, Banana, Papaya, Sapota, Pineapple, Coconut, Cashew, Arecanut, Breadfruit, Jackfruit and Avocado.
2. **Subtropical:** Subtropical climate with distinct winter and summer, eg. Guava, Grape, Citrus, Date palm, Phalsa, Pomegranate, Litchi and Loquat.
3. **Temperate:** Distinct winter, summer and autumn with temperature below freezing during winter is common, eg. Apple, Pear, Peach, Plum, Quince, Apricot, Walnut, Almond, Strawberry and Cherry.

Water

- Water is a transparent, colourless and tasteless liquid compound of hydrogen and oxygen (H_20) with 11.91% hydrogen and 88.81% oxygen.
- It is essential for plant growth and development as a substrate in photosynthesis, regulation of plant temperature, distribution of metabolites and nutrients.
- It comes through precipitation of rain and snow.
- Near equator the total rainfall is 2000 mm per year and away from it, which reduces but again influenced by a number of factors like mountain ranges.
- Water requirement of plant is dependent on soil type and evapo-transpiration rate.
- For crop production it is not the total rainfall but its distribution is more important and in Indian subcontinent we have rains mainly confined to June to September, thereby fruit culture in India had to be supported by irrigation or one has to select crop where fruiting is confined to water availability periods and trees remain dormant during stress.
- Water is also present in the atmosphere as vapour and we call it as humidity.
- This atmospheric humidity also influences growth and development of plants.
- Low humidity has drying effects and enhances water requirement.

- Whereas high humidity favours fungal diseases. Plants liking for high humidity and low humidity are there:
- **High humidity:** Sapota, Banana, Mangosteen, Jackfruit and Breadfruit.
- **Low humidity (Dry)**: Ber, Grape, Date palm, Pomegranate, Citrus, Aonla and Guava.

Air

- A mixture of oxygen, nitrogen and other gases that surrounds the earth and forms its atmosphere.
- It is also one of the climatic factors influencing plant growth.
- If its quality is polluted by the accumulation of gasses like hydrocarbons, SO_2, CO_2, CO, NO, ethylene and methane the plant growth adversely affected but we are more concerned with the movement of air (wind) causing great damage to crops in deserts, coastal areas, valleys for which provision of windbreaks and shelterbelts are suggested and such situations sometimes have to be avoided for plantation.
- Storm has a wind speed of 50/hr whereas, hurricane has a wind speed of more than 100km/hr.

10

Water Management

Water: Water is one of the most important inputs essential for the production of crops. Plants need it continuously during their life and in huge quantities. It profoundly influences photosynthesis, respiration, absorption, translocation and utilization of mineral nutrients etc. Both its shortage and excess affects the growth and development of a plant directly and consequently its yield and quality.

Soil needs the application of water to:

- Remove stress condition.
- Release nutrients in the soil solution for absorption by plants.
- Leach or wash out injurious salts from the soil.
- Preparation of land for raising crops.
- To maintain the temperature and humidity of the soil micro-climate and the activity of soil microbes at optimum level.
- For the normal aeration and functioning of roots and shoots of the plants.
- Excess water needs to be removed for the normal aeration and functioning of roots and shoots of the plants
- Excess water creates unworkable soil condition.

Irrigation: It is defined as the artificial application of water to the plants in the event of shortage of natural rains in order to obtain rapid growth and increased yields. It is an essential item in the cultivation of crops. Success in gardening depends on how efficiently irrigation is provided to gardens because it is governed by many factors such as frequency, duration, intensity, source and method of supply.

Factors affecting the supply of irrigation water to plants

- Topography and soil characteristics.
- Kind of plant (root depth, water absorption capacity, growth habit, etc.).
- Weather condition.

When to irrigate?

- The time when a plant needs irrigation can only be judged by a keen observing eye.
- The plants need water when their new leaves begin to show a wilting appearance. A little before the trees show the sign of wilting.
- The shedding of broad leaves in orchard shows distress symptoms.

How much to irrigate?

- If water supply is limited, only a light irrigation can be given at a time with higher frequency of irrigation.
- If water is available in plenty, the irrigation may be heavy with longer intervals between successive irrigations.
- However, inadequate irrigation reduces the growth and fruiting of the trees while, over irrigation serves no useful purpose and it may even prove to be harmful.
- It may create water logging, the nutrients may get leached and fruits may become watery and develop poor quality.
- Plants which have suffered from drought should not be given liberal doses of irrigations all at once. That may result in the splitting of fruits and even the splitting of bark of the branches and trunk.

Systems of irrigations

- Different systems of irrigation are followed in different parts of the country. The best system is the one which meets the moisture seepage and evaporation.

Principally, irrigation systems can be divided under three broad headings

I. **Surface irrigation:** a. Flooding b. Basin type c. Furrow type d. Ring type

II. **Sub-surface irrigation:** a. Trench method b. Through underground pipelines c. Perforated pipelines.

III. **Overhead or aerial irrigation:** a. Sprinkler b. Revolving nozzles

IV. **Drip or trickle irrigation:**

I. Surface Irrigation

a. Flooding

- When the land is flat, letting in water from one end floods the entire area.
- This system is commonly practiced in canal or tank bed areas.

- It is the easiest method and permits the use of bullock drawn implements in the orchards.
- But in this there is wastage of water and leads to soil erosion also.
- It encourages growth of weeds and spread of diseases like gummosis in citrus and collar rot in papaya.

b. Basin system

- In this system, circular basins are provided around the trunk of the tree.
- The basins are inter-connected in series and are fed through the main channel running perpendicular to the tree rows.
- When compared to flooding, this system minimises the loss of water.
- In this system of irrigation, the water close to trunk may bring about certain diseases like gummosis and nutrients are likely to be carried over from one basin to the other.

c. Furrow system

- Unlike the flood system, here the entire land surface is not covered with irrigation water.
- The furrows are opened in the entire orchard at 4 or less apart, depending upon the age of the trees.
- Water is let in these furrows from the main channels.
- In orchards, two furrows on each side of the rows are generally made.
- It is suited to such lands, which have a moderate slope to the extent of 1-2% if the water is to run freely and reach the ends of the furrows.
- Where the slope is sharp, the furrows are made to follow the contour more or less closely.
- This method has disadvantage of excess of water penetration at the head than at the farther end, which may result in variation in vigour and growth of trees.

d. Ring system

- This is an improvement over the basin system.
- In this system, a ring is formed close and around the tree and water is let into the basin.
- This method is recommended for citrus trees thereby reducing the chances of collar rot to which these trees are often susceptible.

- The size of the ring will increase as the tree grows.
- In this system, the spread of diseases like collar rot, etc., are prevented.
- However, it involves more labour and capital and it does not permit uniform distribution of water throughout the bed or basin as in the basin system of irrigation.

II. Sub-Surface Irrigation

- This system consists of conducting water in number of furrows or ditches underground in perforated pipelines until sufficient water is taken into the soil so as to retain the water table near the root zone.
- In limited situation, this may be a very desirable system of irrigation.
- In general, however, it must be used with great caution because of the danger of water logging and salt accumulation.
- If the sub-strata are so slowly permeable that practically no water moves through, water added may stand in soil sufficiently for long time which results an injury to the plant root due to poor aeration.
- Where irrigation water or the sub-soil contains appreciable amount of salt, sub-soil irrigation is usually not advisable.
- Land must be carefully leveled for successful subsoil irrigation so that raising the water table will wet all parts of the field equally.

III. Over Head or Aerial Irrigation

- In this system, water is applied in the form of spring, somewhat resembling rainfall.
- This is accomplished by pumping water from original source into the main supply line from where it is distributed to perforated pipes, which operate at low pressure (80 to 120 lb per square inch) and supply the water in a fairly uniform rectangular pattern.
- They have a high rate of application, usually 1 /hour or higher. Because of the high application rates, their use is restricted to soils with high infiltration rates, such as sandy or gravelly.
- Revolving nozzle is also at times used, which operated on either low or high pressure. Usually the rate of application followed in the rate of 0.2 to 0.3 per hour.

Sprinkler irrigation

- May have definite economic advantages in developing new land that has never been irrigated, particularly where the land is rough or the soil is too much porous, shallow or highly erodable.

- It is quite useful where only small streams are available, such as irrigation wells of small capacity.
- It is helpful in irrigating at the seedling stage when the furrowing is difficult and flooding leads to crusting of soil.
- Fertilizer materials may be evenly applied by this method.
- This is usually done by drawing liquid fertilizer solutions slowly into the pipe.
- It has several disadvantages like
- High initial cost,
- Difficult to work in windy location,
- Trouble from clogging of nozzle,
- Interference in pollination process
- Requirement of more labours while removing or resetting.
- In general, this system is best adopted for areas where ordinary surface systems are inefficient.

IV. Drip or Trickle System

- This is the most recent system of irrigating the plants.
- It is usually practice for high value crops, especially in green houses and glass houses.
- There will be an installation of pipelines with nozzles very close to the soil.
- The nozzle is fitted in such a way that water is dripped almost in the root-zone of the plants.
- Water is allowed to move in pipes under very low or no pressure and it drop at regular interval.
- This system of irrigation has advantages like no disturbance of the soil; soil moisture is maintained, lesser leaching of nutrients from the soil.

11

Digging and Filling

Digging and Filling of Pits

1. Marking of pits and planting should always be done with the help of planting board.
2. The guide pegs are installed at both the ends.
3. One meter deep pits of one meter diameter should be dug.
4. Top 30 cm soil should be kept on one side, which is used for refilling the pits as it is fertile soil.
5. Bottom 70 cm soil should be kept on other side, which is discarded.
6. The pits should be left exposed for a few days before actual planting.
7. These pits should be refilled with mixture of topsoil, 2-3 baskets of silt and 2-3 baskets of well-decomposed farmyard manure.
8. The refilled pits should be watered a few days before planting the tree. To each pit add carbofuron (25g) for control of white ants.

Method of planting

- Bore holes of suitable size are made in the centre of the filled pits with the help of planting board.
- Place the earth ball of a plant in it in such a way that the upper surface has the same level as ground.
- Fill loose earth around the ball and press it firmly with the handle of a spade or khurpi.
- Apply water soon after planting the plants in the pits.

Planting distance of fruit plants

- The spacing given to the fruit trees is generally governed by the different factors like climate and soil, choice of varieties, growth habit of tree, rootstock used, nature of irrigation and pruning technique followed.

- The spacing may vary according to different systems of planting.
- Provision of optimum spacing to fruit trees is one of the most important aspects of successful fruit culture.
- Optimum spacing regulates the proper utilization of sunlight, avoids competition in the uptake of nutrients caused by the collision of root systems and facilitates proper irrigation.
- The latest technology on high-density plantation system where trees are planted at critical spacing for maximum utilization of space is becoming popular.
- It will be very difficult to suggest exact spacing for fruit trees, which will suit every locality or soil.
- Given below is the spacing of some of the important fruit plants, which serve as basic guideline for establishing a new orchard.

Name of fruit tree	Planting distance (m)	Number of plants/ha (square system)
1. Mango	10	100
2. Citrus & Pomegranate	6	275
3. Grape		
i) Head system	2.0x1.5	3300
ii) Kniffin system	4.00x3.00	1100
iii) Bower system	3.1x6.0	550
4. Guava, Peach & Loquat	6.5	225
5. Litchi & Sapota	7.5-9.0	180-123
6. Ber & Pear	7.5	180
7. Date palm & Almond	6-7	275-202
8. Jamun	10-12	105-75
9. Phalsa & Papaya	1.5	4400
10. Banana		
i) Tall varieties	2.7x3.0	1210
ii) Dwarf varieties	1.8x1.8	3052
11. Pineapple	30x60x90 cm	43500
12. Custard apple	5	390
13. Jack fruit	10	100

Planting season: The planting season of different fruit crops vary on the basis of their evergreen or deciduous nature.

Evergreen fruit plants

- There are two planting times for evergreen fruits i.e., beginning of monsoon and end of rainy season in heavy rainfall areas.

- The fruit plants such as citrus, mango and litchi should preferably be planted during September or in the beginning of October when the weather becomes mild and more favourable and there is enough moisture in the soil.
- Guava can also be planted bare-rooted during February-March or August-September.
- The plant should be defoliated and the roots covered with moist material.
- Most of the sub-tropical and tropical evergreen fruit plants are suitably transplanted during the rainy season when the atmospheric humidity is high and sufficient supply of soil moisture is obtained.
- During this active period of growth, the plants easily penetrate their roots in the soil and get established.
- High atmospheric humidity during the rainy season helps them to minimize the transpiration loss.
- The evergreen plants can also be transplanted during the onset of spring with equal success provided an ample quantity of irrigation water is available.
- Early regeneration of rootlets due to high temperature and available moisture during this period helps in early establishment of the plants in most of the cases.

Deciduous fruit plants

- The deciduous fruit plants are planted during winter when they are dormant.
- Their planting must be completed before the start of new growth i.e., up to middle January in case of peach and plum and up to middle of February in case of pear and grapes.
- The bare rooted ber can also be planted during January and up to middle of February.
- It would be wise, if the planting operation in deciduous fruit plants could be done well before the dormancy is broken and the plants start their growth afresh during subsequent period.
- Planting should be avoided during hot and dry spells of weather.
- The fruit trees should preferably be planted in the afternoon and on cloudy and humid days rather than in bright sunshine and dry weather.

12

Training and Pruning

Horticultral plants are grown for their produce like fruit, vegetable, flower, medicinal, aromatic, spices etc. Therefore, these plants should be managed in such a way that human desires for the purpose of growing them are fully satisfied in terms of quality and quantity of produce. This demands direct manipulation of plant growth itself or plant environment through various inputs. In manipulation of plant development, training and pruning are important for which our knowledge about plant development and its phenology has to be complete. These practices are important in fruit crops.

Training: Physical techniques that control the shape, size and direction of plant growth are known as training or in other words training in effect is orientation of plant in space through techniques like tying, fastening, staking, supporting over a trellis or pergola in a certain fashion or pruning of some parts.

Objectives

- To improve appearance and usefulness of plant/tree through providing different shapes and securing balanced distribution.
- To ease cultural practices including inter-cultivation, plant protection and harvesting.
- To improve performance like planting at an angle of 45° and horizontal orientation of branches make them fruiting better.

Methods of training: Method of training of a plant is determined by the nature of plant, climate, purpose of growing, planting method, mechanization, etc. and therefore, intelligent choice is necessary.

Training in herbaceous annuals and biennials

These plants are usually grown without any attempt to alter their growth patterns because even if useful not practical being in large number in field. However, for some of ornamental value and creeping nature following types of training is affected.

1. Staking or supporting of vine like plants.
2. Training on pergola or trellis of vine type fruit plants or even indeterminate type tomatoes.
3. Nipping of apices for encouraging lateral growth to give bushy appearance or fulsome appearance in pot plants like aster, marigold and chrysanthemum.
4. De-shooting or removal of lateral buds for making single stem for large flowers as in chrysanthemum and dahlia.
5. Staking with bamboo sticks and tying together various shoots in potted chrysanthemum.

Training of woody perennials

The woody perennials, which are widely spaced and remain on a place for a long duration, are trained for develop strong framework for sustainable production of quality produce and for ornamental beauty in different shapes (topiary). In these plants following types of training are followed.

i. Open centre system (vase shaped)

- In this system the main stem is allowed to grow to a certain height and the leader is cut to encourage lateral scaffold from near the ground giving a vase shaped plant. This is common in peaches, apricots and ber (Fig. 7.1).

Fig. 7.1 : Open centre system

ii. Central leader system (closed centre)

- In this system the central axis of plant is allowed to grow unhindered permitting branches all around.
- This system is also known as **closed centre system** and common in use in apple, pear, mango and sapota (Fig.7.2).

Fig. 7.2 : Central leader system

iii. Modified leader system

- This system is in between open centre and central leader system wherein central axis is allowed to grow unhindered up-to 4-5 years and then the central stem is headed back and laterals are permitted.
- It is common in apple, pear, cherry, plum, guava (Fig.7.3).

Fig. 7.3 : Modified leader system

iv. Cordon system

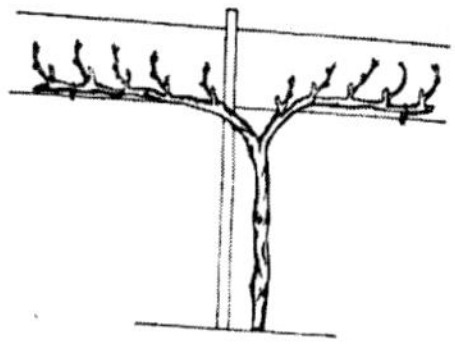

Fig. 7.4 : Cordon system

- This is a system wherein espalier is allowed with the help of training on wires.
- This system is followed in vines incapable of standing on their stem.
- This can be trained in single cordon or double cordon and commonly followed in crops like grape and passion fruit (Fig. 7.4).

v. Training on pergola

- To support perennial vine crops pergola is developed by a network of criss-cross wires supported by RCC/angle iron poles on which vines are trained.
- This is common for crops like grape, passion fruit, small gourd, pointed gourd and even peaches.

vi. Training in different shapes

- Generally ornamental bushes are trained in different shapes for the purpose of enhancing beauty of places.
- These shapes could be vase, cone, cylindrical and rectangular box, flat and trapezoid.
- Presently for the convenience of mechanization these shapes are being utilized in fruit trees.
- Such shapes are given to adjust the geometry of plantation like hedge row system, box, and unclipped natural in fruits like guava, mango, sapota and citrus.

Details of training

1. **Height of the head**: This is the height from ground to first branching or scaffolding.

   Depending on the height the trees could be divided in three groups.

a. **Low head**: 0.7-0.9 m. This is common in windy areas. Such plants are easy to maintain.

b. **Medium head**: 0.9-1.2 m. This is the most common height which combines both effects, ability to stand against wind and easy management.

c. **High head:** More than 1.2 m. Common in tropics in wind free areas. Operations under the canopy are easy to perform.

2. **Number of scaffold branches:** It refers to allowing of number of scaffolds on the primary axis of the tree which vary from 2 to 15 but extremes are undesirable. In fruit trees 5 to 8 scaffolds are preferred to make the tree mechanically strong and open enough to facilitate cultural operations.
3. **Distribution of scaffolds**: Scaffolds should be distributed in all the directions spaced at 45-60 cm allowing strong crotches through wide angles of emergence.

A well trained tree is an asset to the farmer and therefore, efforts should be made for training trees appropriately in formative years for sustainable production. In fact the process should have begun from nursery itself.

Pruning

It is Judicious removal of plant part is known as pruning. It refers to removal of plant part like bud, shoot, root etc to strike a balance between vegetative growth and production. This may also be done to adjust fruit load on the tree.

Objectives

1. To maintain the growth and vigour of the trees and to have a balance between the vegetative vigour and fruitfulness, so as to be conductive for production of optimum crop of best quality.
2. To shape the tree to make the best use of the space between trees while allowing the necessary access.
3. To regulate the size and quality of the fruits by way of proper distribution of the fruiting area.
4. To regulate the succession of crop and to have the crop where it can be managed easily and cheaply.
5. To spread the trees for economic orchard management.
6. To remove the dead, diseased and over aged wood.
7. For effective spraying of pesticides to the crop.
8. To minimize biannual bearing and consequent risk of die back.
9. To get maximum plageiotrophic shoots/stems.
10. Establishment of transplant where leaves/shoots are pruned to strike a balance between roots and shoot so that plants lose less water against restricted root system lost during lifting of plants.
11. Elimination of non-productive vegetative growth like water sprouts, suckers, dead and diseased wood.
12. In case of forest trees production of knot free timber.

Types of pruning

Basically there are three types of pruning with definite purposes.

i. Frame pruning.

ii. Maintenance pruning.

iii. Renewal pruning.

1. **Frame pruning**: This pruning is done to provide shape and form to a plant in its formative years so that tree develops strong framework and a shape for ease of operations. This process begins from nursery itself and continues up to fruiting stage. This is done continuously irrespective of the season.
2. **Maintenance pruning**: To maintain status in production level and for uniform performance this pruning is done. In some plants like grapes, apple, pear, peach etc. (deciduous trees) it is an annual feature and in others (evergreen like mango, sapota) it is rare confining to removal of water sprouts and unproductive growth and opening of the tree.
3. **Renewal pruning**: This pruning is done in old trees like mangoes which shows decline. In this case severe pruning is required.

Factors to be considered in pruning

- In some of the tree species pruning as a regular feature in bearing trees is done to strike a balance between vegetative growth and production so that farmers get sustained production uniformly with optimum quality of produce.
- To achieve this one should consider the following factors.
- Time at which buds are differentiated in relation to blooming.
- The age of the wood that produces the most abundant and highest quality of fruit buds.
- In consideration of these factors our knowledge about bearing habit of the tree/plant should be complete.
- Bearing habit means relative position of a fruit with reference to its potential bud giving rise to flower or inflorescence in the shoot. This habit varies from plant to plant.

Principles of pruning

1. Excessive pruning should be avoided as it affects the growth of the plant by dwarfening and may induce more of water suckers, fasciations (union of a number of parts side by side in a flat plane) and thus affect the bearing potential.

2. In pruning, only that wood which is not necessary for the tree should be removed.
3. Pruning of larger limbs should be avoided as far as possible.
4. Pruning of young trees should be done more carefully than the yielding trees, since severe pruning of young trees delays the cropping and much more of yield area will be removed than what is desired.

Top Working

It is a technique or method of rejuvenation where in the objective is to upgrade seedling plantations of inferior varieties with superior commercial cultivars or hybrids suitable for domestic or export market or the desired variety of the grower. The technique involves grafting with procured scions of desired variety on shoots emerged on pruned branches by adopting softwood grafting during monsoon season (Season of top working slightly varies from species as it also depends on availability of good shoot and scions). The scion shoots and the emerged shoots should be of same thickness.

Advantages of top working

1. Increase the tree productivity /orchard productivity.
2. Conversion of old and senile orchards into productive orchards.
3. Conversion of seedling or inferior variety plantation /orchard into new orchard with desirable variety or varieties through top working.
4. Possibility of grafting several varieties on the same plant.
5. Increasing the fruit set of orchard by grafting few shoots with pollinate varieties.
6. Additional income by selling the pruned wood during non bearing season or period.

Disadvantages

1. Chances of death of plant if not done properly or on severe pruning.
2. Need good management post pruning period.
3. Loss of crop for 2-3 years
4. Chances of pest and disease occurrence (stem borer, anthracnose etc.).
5. Needs skilled labour for thinning of shoots, removal of side shoots etc.
6. Top working technique can be successfully followed in crops like mango, sapota, aonla, cashew, guava, tamarind, jackfruit, etc.

13

Plant Growth Regulators

The quantitative increase in plant body such as increase in the length of stem and root, the number of leaves etc., is referred to as plant growth whereas, the qualitative changes such as germination of seed, formation of leaves, flowers and fruits, falling of leaves and fruits is referred as development. Utilization of these substances for proper development of the plant is regulated by certain chemical messengers called plant growth substances or plant growth regulators, which in minute amounts increase or decrease or modifies the physiological process in plants. Plant growth substances are the chemical other than nutrient which required in very small amount but without which plant can not complete their physiology.

The two sets of internal factors, *viz*., nutrition and hormone control the growth and development of the plant. The raw material required for growth is supplied by nutritional factors which include the minerals, organic substances the protein, carbohydrates, etc.

Phytohormones

These are the hormones produced by plants which in low concentrations regulate plant physiological process. These usually move within the plants from a site of production to a site of action.

Plant growth regulators

- These are organic compounds other than nutrients, which in small amounts promote, inhibit or otherwise modify any physiological process in plant.
- It may be defined as any organic compounds which are active at low concentrations (1-10 ml) in promoting, inhibiting or modifying growth and development in plants.
- The naturally occurring (endogenous) growth substances are commonly known as plant hormones, while the synthetic ones are called growth regulators.

Plant hormones

- It is an organic compound synthesized in one part of plant and translocated to other parts, wherein very low concentration causes a physiological response.
- The plant hormones are identified as promoters (auxins, gibberellin, and cytokinins), inhibitors (abscisic acid and ethylene) and other hypothetical growth substances (florigen, death hormone, etc.).

A. Auxins

- Auxin is a Greek word derived from Auxin which means to *increase.*
- It is a generic term for chemicals that typically stimulate cell elongation by loosening cell wall but auxins also influence a wide range of growth and development response.
- The chemical isolation and characterization was done by Kogi *et al.* (1934).
- Auxins are the first identified hormones of which IAA seems to be the major naturally occurring endogenous Auxin in plants and crops.
- Besides IAA, plants contain three other compounds which are structurally similar and elicit many of the same response as that of IAA, Chloro indole acetic acid (CIAA), Phenylacetic acid (PAA), Indole butyric acid (IBA).

Site of Auxin synthesis: Auxins are synthesized in stem tips and in young tissues and move mainly down stem (Basipetal movement) *i.e.,* from shoot tip to root.

Synthetic compounds are classified into five major categories

- Indole acids
- Napthalene acids
- Chlorophenoxy acid
- Picolinic acid.
- Derivatives.

Role of Auxin

1. **Cell division and enlargement:** IAA + GA, example - cambial growth in diameter.
2. **Tissue culture:** Shoot multiplications (IBA and BAP), callus growth (2, 4-D), root multiplication IAA and IBA (1-2 mg).
3. **Breaking dormancy and apical dominance (inhibition of lateral buds):** NAA

4. **Shortening internodes:** Apple trees (NAA) dwarf branch fruit.
5. **Rooting of cutting:** (10-1000 ppm-NAA, IAA, Phenyl acetic acid)
6. **Prevent lodging:** NAA develop woody and erect stem.
7. **Prevent abscission:** premature leaf, fruit and flower fall (NAA, IAA and 2,4-D).
8. **Parthenocarpic fruit:** Grapes, Banana and Orange (IAA).
9. **Flower initiations:** Pineapple uniform flowering and fruit ripening (NAA) and delay flowering (2, 4-D).
10. **Weed eradication:** (2, 4-D).

B. Gibberellins

- It is the active principle isolated from the soil borne fungus *Gibberella fujikuroi*.
- The concentration of GA_3 is usually the highest in immature seeds, reaching up to 18 mg/kg fresh weight in *Phaseolus* species, but it decreases rapidly as the seeds mature.
- In general, roots contain higher amounts of GA_3 than shoots. Gibberellins have also been found effective in overcoming both kinds of dormancy in buds as well as seeds.

Role of Gibberellins

1. **GA:** Synthesis in leaf and induce shoot elongation (IAA + GA_3), by effecting cell elongation or cell division or both.
2. **Enhance metabolic activity:** Mobilization of reserved food material, promote growth and height, increase root activity and kinetin production in root- translocation to growing bud.
3. **Shoot elongation:** GA_3 spray increases height of seedlings.
4. **Delay senescence:** Increase photosynthetic and protein synthesis so decrease abscission.
5. **Increase cambial growth and differentiation:** Induce flower and fruit set (IAA+GA_3).
6. **Dwarf plant (genetically) to normal height:** GA_3.
7. **Promote flowering in long day plants**: Substitute for long day condition and cold treatment (vernalization).
8. **Induction of parthenocarpy in grapes**: Three physiological events: Rachis cell elongation, flower thinning and berry enlargement.
9. Breaking dormancy and leaf expansion.

C. Cytokinins

- First endogenous cytokinin was isolated from maize kernels named as zeatin.
- Germinating seeds, roots, sap streams, developing fruits and tumor tissues are rich in cytokinins.
- Cytokinins imbibed seeds germinate better in dark than unimbibed lettuce seeds.
- Similarly cytokinin together with gibberellins effectively breaks the photodormancy of celery (*Apium graveolens*) seeds.

Synthetic cytokinins are: Kinetin, Benzyladenine and Ethoxy ethyladenine.

Role of cytokinin

1. Cell division, elongation and enlargement.
2. Tissue culture morphogenesis.
3. Induction of flowering and fruit development.
4. Parthenocarpy.
5. Apical dominance overcoming.
6. Breaking dormancy.
7. Delay senescence.
8. Improves N_2 metabolism.

D. Ethylene

- Neljubow (1901) is credited with having identified the active growth regulating component of the illuminating gas as ethylene.
- Ethylene is formed naturally in plants in amounts sufficient to bring about regulatory effect and it might be considered as plant hormones.
- Recently a synthetic chemical known as ethrel, ethephon, chloroethyl phosphonic acid (CEPA) has been reported to release ethylene when applied to plants.

Role of ethylene

1. Breaking dormancy.
2. Induce ripening of fruits.
3. Induce abscission of leaves.
4. Inhibit elongation and lateral bud growth.

Growth retardant: The term growth retarding chemical or growth retardant is that chemical slows cell division and cell elongation of shoot tissues and regulate plant height physiologically without formative effects e.g: AMO 1618, Phosphon-D, CCC, Chloromequat and Alar.

These do not occur naturally in plants and acts in retardation of stem elongation, preventing cell division. Plant growth retardants are defined as synthetic organic chemicals that cause a retardation of cell division steps in pathways of hormone biosynthesis without evoking substantial growth distortions.

Inhibitors: These suppress the growth of plants. There are phenolic inhibitors and synthetic inhibitors and abscisic acid (ABA).

Phenolic inhibitors: E.g. Benzoic acid, Salicylic acid, Coumaric acid and Chlorogenic acid.

Synthetic inhibitors: E.g. Maleic hydrazide, Tri-Iodobenzoic acid (TIBA), SADH etc. An inhibitor from young leaves of *Betula sps.* prevent the growth of apical buds eg. ABA and Dormin.

Role of abscissic acid (ABA): It should appear as 5th growth regulator in the group; add introduction to ABA; take this before growth retardant

- To stop elongation.
- Induce dormancy.
- Delay germination.
- Inhibit growth process.

Add method of preparation of growth regulator formulations

Methods of application

Growth regulators can be applied in different ways like:

- Spraying method.
- Injection of solution into internal tissues.
- Root feeding method.
- Powder form.
- Dipping of cuttings in solution.
- Soaking in dilute aqueous solution.

Various Uses of Plant Growth Regulators

1. **Propagation of plants:** A number of plants are propagated by stem, leaf cutting and by layering. For promotion of rooting, the most commonly utilized hormone is IBA followed by NAA.

   Gibberllic acid causes inhibition of root formation in cutting. Cytokinins also help in quick and profuse root formation in cuttings and layers. By use of auxins, profuse root formation is observed in cuttings of guava, fig, pomegranate, crotons, rose, hibiscus, etc.

2. **Seed germination:** Many seeds have natural dormancy which can be got over by dipping the seeds in auxins. Soaking seeds of French beans and peas in 10-20ppm solution of GA for 12 hours before sowing, significantly improves the yield and quality. Dipping sweet potatoes in 5ppm GA solution for 5minutes before sowing increases sprouting and yield of potatoes.
3. **Control of plant size:** In fruits and vegetables, application of higher doses of nitrogenous fertilizers spraying cycocel (growth retardant), the superfluous growth of leaves is checked. By spraying 10ppm solution of morphactin in potato, the growth of plant is reduced and thereby the size of tubers is increased. The growth retardants are useful in checking the growth of hedges in ornamental gardens there by reducing the cost of trimming the hedges.
4. **Regulation of flowering:** In Pineapple, due to later flowering the fruit get ready in rainy season. This deteriorates the quality of the fruit. This difficulty can be overcome by spraying 5-10 ppm solution of NAA before flowering. Application of 100-200 ppm GA in Dahlia plants induces early flowering. Sometimes, it is necessary to delay flowering eg. Crossing of varieties which do not flower simultaneously. Hence, the crossing becomes difficult.
5. **Control of sex expression:** In number of cucurbits, such as ridge gourd, bitter gourd, watermelon, cucumber and pumpkins which have proportion of male flowers is more than female flowers. For better yield, it is necessary to increase the number of female flowers. This can be achieved by application of auxins which increases the number of female flowers and decreases the number of male flower. The commonly used auxins are NAA and ethrel.
6. **Control of fruit set and growth of fruit:** Spraying NAA, TIBA, and PCPA on flowers increases the fruit set. Dipping of grape bunches (young fruits) in GA solution increases the berry size in Thompson seedless grape.
7. **Control of fruit drop:** In Nagpur Santra, the fruit drop can be controlled by spraying 10-20 ppm NAA or 10 ppm 2,4-D after fruit set. The fruit drop in mango can be controlled by these two auxins.
8. **Thinning of fruits:** Sometimes it is necessary to thin the fruits so as to bring a balance between the supply of nutrients and development of fruit. In such cases spraying with mild solution of ethrel or morphactin reduces the fruit load by 25-30 per cent.
9. **Early ripening and development of fruit colour:** If the fruits could be brought in the market in early part of the season, they fetch good price. Spraying with 2,4,5-T and B-9 hastens maturity of apples by 1-4weeks.

10. **Prevention of sprouting:** In potatoes and onions, after harvest, in storage, the buds start sprouting which makes them unfit for cooking. Spraying of malic hydrazide (MH) solution before storing prevents sprouting and these can be stored safely for 6 months.

11. **Control of weeds:** The conventional method of controlling the weeds is to remove them by uprooting manually. Successful control of weeds is obtained by spraying 2,4-D in many crops.

14

Bearing Habits

Growth, Fruiting Habits and Methods for Inducing Fruitfulness in Fruit Crops

- Every plant has two distinct phases of development namely vegetative and reproductive.
- In the vegetative phase, the plant only vegetative growth i.e. leaves, shoots etc.,
- In the reproductive phases the plant produces reproductive growth i.e. flowers and fruits.
- In case of annuals, these phases occur for a short period, each phase occurring only once and that too within the same season.
- In perennials, the vegetative phase lasts for several years depending upon the variety and species. This is known as the pre-bearing period.
- Once the perennial tree completes its pre-bearing period, the cycle of vegetative and reproductive phases might occur in each season or the reproductive phase may occur simultaneously along with vegetative phase without much distinction between the two.
- In certain trees the vegetative and reproductive growth may occur in alternate years which are known as alternate bearing.
- In some other trees, the vegetative growth may occur for two, three or more years followed by reproductive growth for one year and again vegetative growth for two or more years. This is known as irregular bearing.
- These buds are normally found towards the apex of the shoot and the axils of the leaves (single bud).
- But in some cases more than one bud is found in the leaf axils as in the case of peach, Japanese plum, grape etc. The extra buds are normally called as supernumerary buds or accessory buds.
- The one which produces floral parts only or a shoot with flowers is known as flower bud, fruit bud or the blossom bud.

- In case the bud produces only floral parts, it is known as a pure flower bud and if it produces a shoot with flowers and leaves it is known as mixed bud.
- The physiological and morphological changes that occur in a vegetative bud in its preparation to change over to the reproductive phase or to become a flower bud may be called as the flower bud initiation.

Bearing Habits

If one observes a mango tree, he/she may find it does produce the fruits terminally on a shoot and it citrus he/she finds the fruits always produced laterally in the leaf axils. The manner or pattern in which a plant produces its fruit is known as bearing habit or fruiting habit. Bearing habits are specific to the kind of plant. In determining the bearing habit of any plant three points are to be considered.

1. The origin or position of the flower or fruit bud which may be either terminal or lateral on a shoot.
2. The nature of the bud, which may be either pure flower bud or a mixed bud.
3. Position of flowers or fruits on the flowering shoots especially in case of mixed bud. The flowering shoot may end with the fruits or carry fruits laterally on it.

Considering the above three points, the bearing habit of fruit plants are broadly classified into six groups.

Group I: Fruit bud terminal, pure flower bud producing only flower parts, e.g., mango, cashew.

Group II: Fruit bud terminal, mixed bud with flower parts situated terminally on the flowering shoot, e.g., apple, pear.

Group III: Fruit bud terminal, mixed bud, with flower parts situated laterally on the flowering shoot, e.g., guava.

Group IV: Fruit bud lateral, pure bud producing only flower parts, e.g., citrus, coconut, date, gooseberry.

Group V: Fruit and lateral, mixed bud, with flower parts terminally on the flowering shoot, e.g., grapes, cashew and Annonaceous fruits.

Group VI: Fruit bud lateral, mixed bud, with flower parts laterally on the flowering shoot e.g. fig, mulberry, ber etc.

- Of these six groups, the first three come under terminal bears, through we see fruits laterally on the flowering shoot in Group III.
- The other three groups are lateral bearers through we see fruits terminally on the flowering shoot in group.

- In addition to these six main groups we do find some different bearing habits in some plants which cannot be conveniently classified under any of these six groups.
- These are therefore classified into separate groups.

  Group VII: Flower buds borne sub terminally in the leaf axils, pure flower bud. This could be treated as a lateral bearer e.g. papaya and sapota.

  Group VIII: Flower buds borne adventitiously on the main trunks or branches, pure flower bud, e.g. jack fruit and cacao.

  Group IX: The growing apex (pseudo stem) ending in a flower which could be treated as a terminal bearer e.g. banana and pineapple.

  Group X: Fruit bud terminal or lateral, pure bud. These plants come under group I and IV and as such classified separately e.g. pomegranate.
- In case of these groups occasionally the plants may bear in different pattern also deviating from its own characteristic pattern but majority of its bearing will be always as per the classification.
- The knowledge of bearing habits will be of much importance for regulating the bearing of any plant by means of pruning, terminal bearers are not normally pruned.
- Whereas in case of lateral bearers pruning certainly increase the bearing, as it encourages the sprouting of lateral buds.

Methods for Inducing Fruitfulness

The fruitfulness of a tree is governed by various external and internal factors. The most important external factors are environment, insect pests, disease and nutrient supply to the plants. The important internal factors are sex distribution, heterostyly, dichogamy and aborted organs, non- viability of pollen, genetic incompatibility and nutritional status of the plant.

External factors: The environment comprises of temperature, rainfall, wind and light.

1. **Temperature** plays on important role in determining the fruitfulness.

- Chilling is required for breaking dormancy and lack of it during winter is responsible for low yields of an apple at Connor.
- Pollen of many deciduous fruit plants may germinate very freely at optimum temperature.

2. **Light intensity** also influence on fruit set to a certain extent

- Interior branches of a tree, which bear fewer fruits.

- In strawberry the flowers developed stamens when they are exposed to long daylight.

3. **Rainfall** also determines the extent of fruitfulness.

- The period of water stress appears to be a most potential factor for influencing flowering in tropical fruit trees.

4. **Infestation of insect pests and diseases** also reduced the fruit set considerably.

- The best examples are mango-hoppers, malformation of mango inflorescence and anthracnose in grapes.
- By preventing the attack of these pests and diseases by suitable control measures in the beginning itself, one can actually increase the fruitfulness of these trees.

Internal factors

- In some of the fruit plants, individual plants are self-unfruitful as their flowers are unisexual i.e., flowers of one sex may be present in one plant as in case of papaya, strawberry, fig and date palm.
- The fruitfulness of these plants can be improved by inter planting of the plant bearing opposite sex flowers.
- Prevention of self pollination in perfect flowers may be due to the difference in the time of maturing of male and female sex organs and this is known as dichogamy, as found in Avocado.
- Genetic incompatibility is another cause of unfruitfulness. The pollen of some varieties are not capable of fertilizing the flowers of the same variety or certain other varieties as in the case of apple, sweet cherry, plum, pear, almond, mango, sapota etc, in such plants to improve the fruitfulness, certain suitable combination of varieties which would successfully cross pollinate each other should be planted.
- Fruitfulness of a tree is to a great extend determined by the nutrient supply and the nutritional status of the plant.
    - Application of manures and fertilizers a few days before blossom emergence is generally believed to increase the fruitfulness.
    - Relationship between C/N ratio and fruitfulness.
    - The fruit bud formation, setting of the fruit and its development mainly depends upon the requisite balance of nutrients in the branches of the trees at certain critical seasons.
    - Plants with plenty of nitrogen and high carbohydrate accumulation are found to make moderate growth and produce satisfactory crop. Whereas,

plants with plenty of nitrogen and moderate carbohydrates are found to grow vegetative, the expense of fruit production. By adopting some of the horticultural practices like root exposure, root pruning, bending, girdling or ringing, notching and smudging, the plants can be made to accumulate more of carbohydrates at the desired positions of the plant and these make them more fruitful.

1. Root pruning

- This is a recognized practice in the dwarf fruit tree culture.
- Every year a trench is dug a few centimeters away from the trench of the previous year.
- The trench is then filled with manures liberally and watered.
- As a result of this, circular mass of fibrous roots increases very slowly from year to year and the tree becomes short but stocky and well matured shoots will be rich in their reserve food material and will be more fruitful.

2. Root exposure

This is commonly practiced to force flowering and fruiting of orange, guava and other citrus fruits in the desired season. It is called as Bahar Treatment.

- About two months before the bloom, the soil around the tee is removed near the main roots from an area of 60cm radius.
- The fibrous roots are removed and the main roots are exposed to the sun.
- The trees are allowed to go dry until the leaves wither or even some of them fall. Time taken for leaf fall is 3-4 weeks.
- After this stage, the exposed roots are covered with a mixture of soil and manure and watered immediately.
- During this period of root exposure, the trees are forced to take rest since the upward movement of water is adversely affected.
- Within three weeks of covering the roots and watering, the trees burst into heavier bloom, and set a much larger crop.
- This treatment can be practiced only in orchards planted in retentive soils which remain moist for a long time and do not allow plants to take a post harvest rest even though watering is stopped.
- Now days this treatment is becoming up popular in view of the fear that it may adversely affect the performance of the trees in the long run.

3. Bending

- This is widely practiced for increasing the fruit production in guava, especially in the erect growing varieties.

- The large branches in their erect and upright position have a natural tendency to produce fruit bearing shoots near about their ends, while the lower branches remain more or less dormant.
- When such large branches are bent and tied and pegs fixed in the ground in an arch fashioned way, the fruiting area is considerably increased due to the increased number of side branches produced on the lower parts of the main limbs.
- This practice is not necessary in spreading varieties like Sardar guava (Lucknow- 49)

4. Ringing or girdling

- This is one of the well known methods of increasing fruit bud formation.
- The operation consists of removing a strip of bark of a tree.
- It interrupts the downward movement of carbohydrates and thus causes them to accumulate above the ring or girdle.
- In India girdling is sometimes practices on mango to force flowering in over vegetative trees, which normally do not bear satisfactory crops.
- The branches of 15 to 20 cm thick are girdled by removing a strip of bark about 2 cm wide all around at the base of the branch or little above the point where it joins another branch.
- In santra, saw passed round the branches is just enough to cause the ring.
- It is usually done between 15th of April and 15th of May.
- The healing of the ring and complete restoration of new bark on the girdled portion is highly imperative after fruits have set.
- The ring of bark removed should be such as not to injure the cambium.
- It is recommended to practice the girdling on the trees which are vigorous in growth. The exhausted trees should be avoided.

5. Notching

- This is practiced in fig. This consists of removal of the small narrow strip of bark just above a dormant bud.

  Notching affects the sprouting in three ways.

  - Firstly, it prevents the inhibitory influence of certain compounds in the bud.
  - Secondly, it cuts off the supply of carbohydrates synthesized from the leaves above and increased the supply of nutrients from below.

- Thirdly, the bud which was once one of the lower most of the branches becomes apparently the top most due to construction in the bark connection with the part of the branch above the notch. Notching can be given below bud also.

6. Smudging

- This is practiced for mango in Philippines to produce off season crop.
- This consists of burning trash wood on the ground and allowing the smoke to pass through the crown of the tree.
- To direct the smoke to different parts of the tree, a cone shaped enclosure with a tall chimney at the top is built with wooden stakes and thin bamboo strips woven together.
- The tree will be smoked heavily and continuously for a week and then mild fire are made at regular intervals for about a month or till the tree comes into bloom or till the terminal buds start swelling and show the symptoms of sprouting.
- Smoke is known to contain ethylene, which is a flowering hormone.

Alternate Bearing

Tendency of variety to bear heavy crop load in one year (on year) and reduce crop load in following year (off year) is known as alternate bearing.

Factors affecting the alternate bearing: The following factors are affecting the alternat bearing.

1. **General factors:** In general the north Indian varieties are biennial in nature whereas, south Indian verities are regular bearing eg.-Neelam, Banglora and Rumani etc.
2. **Fruiting pattern:** Plant have the two types of fruitinfg pattern i.e. terminal bearing and axillary bearing. The terminal bearing plants are always alternate bearers because its flowers comes always in one year old branchage whereas, the axillary bearing plants are always regular bearer because its flowers comes always in current season growth. The auxiliary bearing plants require regular pruning for pramoting current growth.

   Regarding fruiting patern the two terms are neccesory to understand i.e. Flower Bud Differenciation (FBD) and Flower Bud Initiation (FBI).

   F.B.D: It means the anatomical transition of vegetative primordia to reproductive primordia.

   F.B.I: It means emergence of flowering bud.

When FBD process completed then FBI started. In general the South Indian varities flowers early in comparision to North Indian varities. This is only due to the FBD process comes early in South Indian varities as comparision to North Indian varities

3. **C:N ratio:** In general the C:N ratio is very important. The high carbohydrate and high nitrogen favours FBD whereas, low level of CHO reverse flowering and fruiting. In general there are four type of C:N ratio i.e. CCCC:N (very high carbohydrate and very low nitrogen), CCC:NN (Moderate carbohydrate and low nitrogen), CC:NNN (Low carbohydrate and moderate nitrogen) and C:NNNN (Very low carbohydrate and very high nitrogen). Out of these the moderate carbohydrate and low nitrogen and the moderate nitrogen and low carbohydrate favours the flower bud differentiation.
4. **Hormonal factor:** Beside C:N ratio there are some indigenous substances are also responsible for alternate bearing. Thel level of auxin, GA_3 and ABA are also responsible for flowering and fruiting. The plants have high auxin and low GA_3 induce more flowering whereas, low auxin and high GA_3 reduce the flowering.
5. **Climatic factor:** The climate play and important role in alternate bearing. When temperature is high plant have more flowering because at the time of high temperature the level of ABA is high and the level of GA_3 is low, due to synergistic effect of ABA and ethylene the level of ethylene also high due to high concentration of ethylene the more flowering take place in the plant. At the same time if temperature is low reduces flowering and more number of male flowers. This is due to low temperature level of ABA is low due to antagonistic effect of ABA and GA_3 plant have increase the GA_3 concentation i.e. why vegetative growth takes place in the plant due to high concentration of GA_3 and flowering is low.
6. **Cultural factors:** If orchard is not properly managed then there is very less chance of flowering and fruiting.

Overcome of biennial bearing: For overcome of biennial bearing following point are commonly practices.

A. Proper management of orchard.

B. To maintain efficient C:N ratio.

C. By smudging- (Smudging is the procedure by which temporary inhance the temperature of orchard by fumigation for promoting the flower bud initiation).

D. Application of paclobutrazol@5ml/tree (Paclobutrazol is an antigibberalin chemical, by application of which plant have reduces the level of GA_3 and induce the level of ABA).

E. Always planting regular bearing variety like Mallika, Amrapali, Ratna, Sindhu, Arka Puneet etc.

15

Problems of Unfruitfulness

In an orchard all the fruit trees do not bear equally or regularly and sometimes fail to flower and fruit under similar conditions where another fruit tree bears heavily. This failure to set fruits may be attributed to unfruitfulness. To understand the problem of unfruitfulness in orchards a familiarity with following terms is necessary.

1. **Fruit setting:** It refers to initial growth of ovary and its associated parts after blossoming and taking it to maturity.
2. **Fruitfulness:** It is the state of plant when it is not only capable of flowering and fruit setting but also takes these fruits to maturity and inability to do so is unfruitfulness or barrenness.
3. **Infertility:** Ability of a plant not only to produce fruits but develop viable seeds and the inability to do so is referred as sterility or infertility. All fertile plants are fruitful but all fruitful plants are not fertile (Seedless fruits).
4. **Self fruitfulness:** Ability of a plant to mature fruits after self pollination.
5. **Self fertility:** Capacity of a plant for the production of viable seeds after self pollination.
    - The ability of a plant to produce optimum crop is Fruitfulness.
    - The inability to achieve this is referred to as Unfruitfulness.

This unfruitfulness is one of the serious problems of many orchards and its causes need to be understood properly for effective control and obtaining economically acceptable production level. The causes to this problem can be many and they have been broadly grouped into two categories

A. Internal factors

B. External factors.

A. Internal factors associated with unfruitfulness

There are a number of internal factors which are associated with unfruitfulness or sterility. They have further been categorized into three major categories, they are

1. Evolutionary tendencies.
2. Genetic influence.
3. Physiological factors.

1. Evolutionary Tendencies

In the process of evolution, a number of situations may lead to imperfect flowers or varied developmental periods leading to unfruitfulness unless suitable measures are adopted.

i. Monoecious and Dioecious nature

- A plant with stamens and carpels in different flowers on the same plant is monoecious.

  e.g. Coconut, Arecanut, Pecan nut, Capri fig and Hazel.
- In monoecious fruit plants in general there is no or very little problem of pollination, fruit setting and fruitfulness. Nevertheless, pollinators need to be ensured.
- Plants which bear male and female flowers on different plants are known as Dioecious. **e.g.** Papaya, Date palm and Strawberry.
- Likewise a few varieties of plum produce too little pollen to call them bisexual.
- Profuse flowering without fruit set in ornamental pomegranate is a result of their being unisexual.
- A number of sex forms have been reported in papaya by different scientists.
- In case of figs two types of flower clusters are borne namely staminate and pistillate flowers.
- In Capri fig staminate flowers are borne near the eye and pistillate flowers are borne near the end. To ensure good fruit set, retention of a few staminate trees (9:1) is essential as pollinizers.

ii. Heterostyly

- A condition in the flower where length of the style, relative to other parts of the flower, differs in the flowers of different plants.
- In this case in some flowers styles are short with long filaments and in some of the flowers of some species or varieties styles are long with short filaments.
- Thus styles and stigmas at different height prevent self pollination.
- In case of brinjal there are 4 types of flowers according to their length of style i.e. long, medium, pseudo short and true short. Out of these pseudo short and true short do not produce any fruit.

- Similarly in delicious group of apples extreme upright positions of the stamens accompanied by spreaded petals do not permit bees to do pollination while collecting nectar.
- When the pistils of heterostyled plants are pollinated with pollen from the same flowers or from other flowers containing stamens of an equal height the union may be fruitful but it is likely to be of varying degree of sterility. Here arrangement for cross pollination needs to be created.

iii. Dichogamy

- When stigmatic receptivity period does not coincides with pollen viability in monoecious plants it is known as dichogamy.
- In dichogamy self pollination is prevented in perfect flowered plants, due to maturity of two sex elements at different times.
- If the stamens ripe before the stigmas become receptive the flowers are known as **Protandrous** and if stigmas become receptive before the stamens produce viable pollens it is known as **protogynous**. This results in low production of fruits.
- Protogyny is present in monoecious plants like walnuts, hazels, etc. whereas protandry is present in many coconut varieties.
- Majority of dioecious plants are also protogynous.

iv. Abortive flowers or aborted pistils or ovules

- This occurs in the developing flower's pistils and stigmas of many species and is responsible for failure in fruit setting.
- Abortion of partially developed flower buds is common. Setting and maturity of two sexes depend on the erosion of two properly formed sex cells.
- Any interference with their development and functioning may lead to sterility or unfruitfulness; such things can be observed in some grape varieties and tomato varieties.
- The late flowers of strawberry cluster are always abortive. This is more common in indeterminate type of plants.
- Degeneration of pistils takes the form of abortion and it is more common in ornamental pomegranate.
- Certain olive varieties have 10-60% abortive embryos.
- It is also common in some apple varieties. Embryo sac abortion becomes a cause of seedlessness in certain instances than fruitfulness.

v. Impotence of pollen

- Many varieties of grapes produce non viable or impotent pollens though they appear as perfect flowers.
- Sterility in grape varieties was the result of impotent pollens. Sterile pollen in the grape results from degeneration processes in the generative nucleus or arrested development prior to mitosis in the microspore nucleus.
- This is also common in J.H. Hale peach, Washington Navel orange and Tahiti lime.

2. Genetic Influences

- Self sterility is a condition determined by the inheritance received but can develop in favourable environment. Self sterility affects it's off springs as well as hybrids.

i. Sterility and unfruitfulness due to hybridity

- Generally wider the crossing, greater is the degree of sterility encountered.
- The cross between peach and plum bears abundance of flowers but they are without pistils with malformed stamens.
- Flower characteristics were constant sterile and barren.
- A hybrid between the pear and the quince was seedless.
- Most of the citranges (cross between sweet orange and *Citrus trifoliata*) produce no fertile female gametes.
- Seedlessness in most of the banana and pineapple varieties is due to hybrid nature of their ancestors.
- Most of the triploid apple varieties produce aborted pollen.
- A number of hybrids between *Vitis rotundifolia* and *Euvitis* are completely sterile.
- Similar was the case with hybridization of *Vitis vinifera* and *Vitis rotundifolia*.

ii. Incompatibility

- One of the most common causes of self unfruitfulness and self sterility is due to incompatibility between the pollen and ovules of the same plant or of the same variety.
- Pollen and ovules are fertile but they fail to affect conjugation.
- In apple, pear, plum and aonla self incompatible varieties require another pollinizer varieties for fruit setting.

- Self incompatibility has been reported in some of the mango varieties like Langra, Dashehari and Chausa.
- Self sterility and self unfruitfulness has been reported in apple, pears, plums, almond, apricot, the Clementine' mandarins, may be attributed to incompatibility where normal processes of fertilization fails somewhere between production of functional gametes and the fusion of sex cells.

3. Physiological Influences

i. Slow pollen tube growth

- Slow growth of the pollen tube results in unfruitfulness.
- Differences have been found in the rate of growth in selfed and crossed apples, pears, cherries and certain citrus fruits.
- This may be considered one type of incompatibility due to chemotropic or hormone influences.
- Besides this, fertilization should take place within a short time failing which abscission will take place at the base of the style, ovary pedicel or peduncle and fruit setting does not take place.

ii. Premature or delayed pollination

- Premature or delayed pollination leads to unfruitfulness.
- Tobacco flowers are very susceptible to injury from premature pollination.
- When mature pollen grains are applied to immature pistils they germinate, penetrate the style, enter the ovule and if the ovules are not ready for fertilization the flowers fall.
- However, in case of oranges premature pollination did not have any deleterious effect whereas some injury was noticed in tomato.
- Lower setting due to premature pollination was noticed in persimmon, Pear, plum and peach.
- Similarly, if pollination is delayed the flowers fall without setting.
- Delay in pollination for 1 or 2 days did not affect fruit set. However, further delaying may result into polyembryonic seeds in some species.

iii. Nutritive condition of plant

- Nutritive condition of plant just before or at or and just after the time of blossoming is an important factor determining the percentage of flowers carrying for setting and for maturity.
- It may affect the pollen viability or fertility of pistils.

a. **Effect on pollen viability** : There was significant difference in germination percentage of pollen collected from old apple trees and from strong young trees of the same variety.

b. **Effect on defectiveness of pistils:**

- Exhaustion of tree by over bearing, drought or poverty of soil leads to production of defective pistils.
- Over bearing weakens the fruit tree and in coming season production is adversely affected.
- Close correlation was reported between defective pistils and unfruitfulness in American plums.
- In case of *Vitis vinifera* carbohydrate deficiency is the common cause of flower drop.
- Due to carbohydrate deficiency flower abortion and ultimately unfruitfulness also occur in green house grown tomatoes.

iv. Fruit setting of flowers in different positions

- Fruits borne on terminal growth have more competition in many fruit crops and mature and set under normal nutritional conditions but percentage of set is small.
- This positional competition takes place between fruits and branch as well as between different fruits influencing fruitfulness.

Strong and weak spurs

- Nutritional condition of spurs has positive correlation with fruit setting in apple. Spurs on vigorous limbs with large leaves set more fruits than those borne on weak limbs.
- More flowers ultimately lead to more fruit set and more flowers are generally borne on strong limbs. Likewise flowers borne singly set fruits and mature as fruit and majority of those born in clusters drop down.
- Ringing or girdling also lead to accumulation of an extra store of food material leads to fruit set and develop parthenocarpically.
- In the process of fruitifications the embryo is more important for development *i.e.* if nutritive condition is favourable, it accompanies the development of the seed coat and fruit wall, if not, only the latter portions are in high degree retardation in development.
- Under insufficient nutrient supply the numbers of seed forming ovules are diminished and under extreme nutrition deficiency both fruit wall and large number of ovules are diminished leading to enabling to form seed.

- In case of green house cucumbers, nutritional deficiency leads to arrest of growth of growing fruits depending upon the position of the fruits and time of pollination. If a few of the cucumbers are harvested remaining fruits resume growth.
- In case of strawberries producing bisexual flowers may lead to produce pistillate flowers if nutritional deficiency was observed.
- However, nutritive condition has indirect influence on compatibility.

B. Unfruitfulness associated with external factors

1. **Nutrient supply**

- In certain families like Graminae, Cruciferae and Leguminaceae sterility normally occur due to over feeding.
- Jonathan apple self sterile in rich soil becomes self fertile in poor soils.
- High fertility level is generally associated with good pistil development and low level with poor pistils and good stamens in grapes.
- In olives low fertility leads to partial or complete degeneration of pistils.

2. **Pruning and training**: Pruning tends to produce more true hermophrodite condition in grape variety "Hope". If pruning is not done the variety tends to remain sterile and produces aborted pistils.
3. **Locality**: Jonathan apple which is sterile in one location is reported to be self fertile in another location.
4. **Season:** Hybrid grape Ideal is self impotent early in season but becomes self potent later on.
5. **Temperature**: High temperature at flowering dries up stigmatic secretion and prevents pollination. Tomato varieties grown at high temperature do not produce any fruit.
6. **Light**: Exposure of strawberry plants to long photoperiod results in development of stamens and pistils in strawberry flowers.
7. **Pests and diseases**: Mango hopper, powdery mildew, etc. adversely affect the fruit set and development in mango and grape.

- Spraying the trees when they are in bloom i.e. spraying at flowering reduces fruit set.
- Some of the fungicides gave inhibitory effect on pollen grains i.e. copper fungicides at 200 to 10000 ppm prevent the germination of pollen grains on the stigma.

Steps to Overcome the Problem of Unfruitfulness

- Having known that there could be many reasons for unfruitfulness, it is necessary to make necessary corrective measures which should begin from planning level and extend to an established orchard.
- Choice of the crop and variety should be made on the basis of climatic and edaphic conditions of the site of orcharding.
- Provision of windbreak and shelter belts for areas prone to wind damage.
- In case of problems of pollination due to heterostyly, dichogamy incompatibility, sterility, embryo abortion, hybridity, etc. a mixture of varieties should be grown by introduction effective pollinizer varieties and pollinators (Honey bees).
- Unfruitfulness due to slow growth of pollen tube, premature and delayed pollination, use of plant regulators can be affected after standardization in terms of chemical concentration and timing of application.
- Problem due to overbearing can be managed through thinning at appropriate stage.
- Irrigation management would be key role in situations with drought and waterlogged conditions.
- Maintenance of critical nutrient status in tree leaves for optimum crop production by adopting correct nutritional programme based on plant and soil analysis.
- In crops requiring regular pruning standard practices will have to be adopted based on crop, variety and its phenology.
- Unfruitfulness due to pathogens should be managed through effective plant protection measures following integrated approach.
- Problem of unfruitfulness due to tendency of alternate bearing should be over come through replacement of regular bearing varieties and crop regulation.

16

Pollination

Pollination : Transfer of pollen grains from anther to stigma is termed as pollination.

1. Self pollination

When pollen grains from an anther fallen to stigma of same flower or same plant is called self pollination or autogamy. In other word when pollen and pistil are from the same plant, often (but not always) from the same flower. In other word when transfer of pollen grain from anther to stigma of a same flower or same plant is called as self pollination. In which development of seed by self pollination is known as autogamy it is the closest form of breeding. Autogamy leads to homozygosity such species develop homozygous balance and do not exhibit significant inbreeding depression several mechanisms which promote autogamy.

a. **Bisexuality**- Male and female organs on same flower

b. **Homogamy**- Maturation of anther and stigma of flower at same time eg. Citrus, Peach & Apricot.

c. **Cleistogamy**- Pollination and fertilization occur in unopened flower bud and dehiscence of the pollen grain takes place before opening of the flower.eg. Grape, Sapota

d. **Chasmogamy**- Fertilization after opening of flower

e. **Position of anthers**- Stigma is surrounded by anther in such a way that self pollination is ensured.

f. **Geitonogamy**- Pollination occur in monoecious plant.

Main feature of self pollination

1. The transfer of pollen grain from anther to stigma of same flower (autogamy) or to the stigma of another flower of same plant (geitonogamy).
2. Need no external agency or agent.
3. Male and female parts mature at same time (homogamy).

4. New varieties cannot be produced, new variations are not possible, less vigorous, offsprings are produced and thus cannot adapt to environmental change.

   eg- Bean, Cowpea, Lettuce, Tomato

2. Cross pollination

When pollen and pistil must be from different plants is called as cross pollination. In other word when transfer of pollen grain from anther to stigma of a different plant by any pollination is known as cross pollination. Development of seed by cross pollination is called allogamy. This is common form of out breeding allogamy leads to heterozygosity. There are various mechanisms which promote allogamy.

a. **Dicliny :** It refers to unisexual flower monoecy and dioecy.

b. **Dichogamy** : Maturation of anther and stigma of the same flower in different time i.e. protandry and protogany.

c. **Heterostyly :** Style and filaments in a flower are of different length.

d. **Herkogamy :** Hindrance to self pollination due to some physical barriers such as presence of hyline membrane around the anther.

e. **Self incompatibility** : Inability of fertile pollen to fertilize the same flower.

f. **Male sterility :** Pollen grain are non functional.

g. **Geitenogamy :** When pollen from flower of a plant fall of the stigma of other flower of the same plant or variety known as geitenogamy.

Monoecy: When male and female flowers are separate but present in the same plant. When male and female flowers are present in the same inflorescence such as in mango, castor and banana.

Dioecy: When staminate and pistilate flower are present on different plant It is called dioecy eg- Papaya, date palm.

Main feature of cross pollination

1. The transfer of pollen grain from anther of flowers of a particular plant to the stigma of flowers of another plant if same species (allogamy).
2. Need external agents like birds, insects, winds, water etc.
3. Male and female parts mature at different time (dichogamy).
4. New varieties may be produced by cross pollinating two different verities of same species, new variations are possible as offspring are healthier and thus are adopt themselves to environmental change.
5. It is also called as allogamy, xenogamy etc.

Eg. Amaranths, Beat leaf, Spinach, Garden beet, Sugar beet, all Cucurbits, all Cole crops, Radish, Turnip, Carrot Onion etc. Mango (housefly) Citrus, Papaya (wind) Guava (wind), Sopota, Jack, Litchi and all fruit crops.

Pollinator: Most plants needs help moving pollen from one flower to the pistil of another flower. Wind moves the pollen for same plants such as grasses like corn. Animal pollinators move pollen for many other flowering plants. Animals that are known to be good pollinators of flowers includes bees, butterflies, hummingbirds, moths, some flies, some wasps and nectar feeding bats.

Benefits of pollination: Plants benefit from pollinators because the movement of pollen allows them to reproduce by setting seeds. However, pollinators donts know or care that the plant benefits. They pollinate to get nectar and or pollen from flowers to meet their energy requirements and to produce offspring. In the economy of nature the pollinators provide an important service to flowering plants while the plants pay with food for the pollinators and their offspring.

Fertilization: Fusion of the one of sperm with egg cell, producing a diploid zygote and male gamete with female gamete is known as fertilization. In other word it is also known as generative fertilization insemination, pollination, fecundation, syngamy and impregnation is the fusion of gametes to initiate the development of a new individual organism. The cycle of fertilization and development of new individual is called sexual reproduction. During double fertilization in angiosperm the haploid male gamete combines with two haploid polar nuclei to form a triploid primary endosperm nucleus by the process of vegetative fertilization.

Cross fertilization: It has been shown in the present volume that the offspring from the union of two distinct individuals especially if their progenitors have been subjected to very different condition have an immense advantage in height, weight constitutional vigour and fertility over the self fertilized offspring from one of the same parents and this fact is amply sufficient to account for the development of sexual elements that is for the genesis of the two sexes.

Double fertilization: After pollen is deposited on the stigma it must germinate and grow through the style to reach the ovule. The microspores or the pollen contain two cells the pollen tube cell and the generative cell. The pollen tube cell grows into a pollen tube through which the generative cell travels. The pollen tube is guided by the chemical secreted by the synergids present in the embryo sac and it enters the oval sac through the micropyle of the two sperm cells one sperms fertilizes the egg cell, forming diploid zygote, the other sperm fuses with the two polar nuclei, forming a triploid cell that develops in to the endosperm. Together these two fertilization events in angiosperm are known as double fertilization. After the fertilization is complete, no other sperm can enter.

Difference between pollination and fertilization

- Firstly pollination is the transfer of pollen from anther (male organ of the flower) to the stigma (female organ of the flower) of the same or different flower is called pollination.
- Fertilization occurs once the pollen grain reaches the stigma it produces a pollen tube which grow down through the style to the ovary.
- Pollination is a very of reproduction for flowering plants. The word comes from the word pollen which is basically the sperm of the plant. The pollen makes contact with the plants stigma and then the fertilization process start. The process of pollination was actually discovered in the 18th century by a man named Christian Sprengel.
- Fertilization is the joining of gametes together in order to produce an offspring. The female egg will be fertilized by a male sperm and this will leads to the creation of a child, be it for animals or for the plants. Basically fertilization occurs not only in animals but in plants a well.
- Pollination is a process flowering plants only undergo transfer of pollen to the plants stigma. The process can be done by the plant itself or through outside agents.
- Fertilization is basically the joining of sperm and egg for flowering plants there can be no fertilization if there is no pollination. However fertilization applies to almost every organism in this world.

Pollination and reproduction groups of crop plants: On the basis of mode of pollination, crop plant can be classified into three groups.

1. **Autogamous**: It is also propagated both vegetative and seed propagation (extent of cross pollination {0-5%}) eg. Potato Tomato, Bean, Pea, Lettuce etc.
2. **Allogamous:** It is both also propagated by vegetative and seeds (extent of cross pollination {>12%}) eg. Cucurbits, Root crops, Onion Coffee etc.
3. **Often allogamous**: It is also called as often cross pollinated crops. In this group the cross pollination is found more than 5%. It is also study propagated by seeds crops (extent of cross pollination {5-12%}) eg. Lima bean, Egg plant, Okra, Chilli, Sweet pepper.

1. **Protandry:** In which androecium mature first than gynoecium eg. Walnut and Coconut etc.
2. **Protogyny:** Protogyny where gynoecium mature first than androecium eg. Annona, Banana, Plum, Pomegranate, Avocado, Fig etc.

17

Post Harvest Technology

Importance of Post Harvest Technology

1. Proper handling, packaging, transportation and storage reduce the post harvest losses of fruit and vegetables. For every one percent reduction in loss will save 5 million tons of fruit and vegetable per year.
2. Processing and preservation technology helps to save excess fruit and vegetable during the glut season (off season).
3. The technology has become a necessity to improve the food safety and strengthen nation's food security.
4. The technology helps to boost export of agricultural commodities in the form of preserved and value added products.
5. Presently mango, pineapple, citrus, grapes, tomatoes, peas, potato and cucumber being processed on a large scale.

Principles and Methods of Preservation of Fruits

Preservation: Preservation means just protect the foods against the spoilage, but scientifically it may be defined as a science which deals with the process for prevention of decay or spoilage of the food is called preservation. In other words, just controlling the physical, chemical or microbial changes in the foods is called preservation.

1. **Physical changes**: Colour, flavour, texture and taste *etc.*
2. **Chemical changes**: Carbohydrate, fats, proteins, vitamins and minerals.
3. **Microbial changes**: Mould, yeasts and bacteria

Why do we preserve the food?

1. To supply to increase the shelf life of the food for increasing the supply.
2. To make the seasonal fruits available throughout the year.
3. To add the variety to the diet.
4. To save time by reducing preparation, time and energy by fire.

5. To stabilize the prices of the food in the market.
6. To improve the health of the population.

Principles of preservation: There are three main principles:

A. Prevention/delay the microbial decomposition of the food.
B. Prevention/delay the shelf decomposition of the food.
C. Prevention of damage by insects, animals, mechanical causes *etc.*

A. Prevention/delay the microbial decomposition of the food

1. By Keeping out the micro organisms- Asepsis
2. By Removal of micro organisms- Filtration
3. By Hindering the growth and activity of micro organisms- Anaerobic condition
4. By killing the micro organisms- Exposing at high temperature

    1. **Asepsis:** It means preventing the entry of micro organisms by maintaining of general cleanliness, while picking, grading, packing and transporting of fruits and vegetables, increase their keeping quality and the product prepared from them will be superior quality.
    2. **Filtration:** Fruits juice, bear, soft drinks, wines *etc.* enter through bacteria a proof filter which is made of Asbestos pad or unglazed porcelain type of materials. These filters contain the micro organisms and allow the water or juice to percolate though with or without pressure.
    3. **Anaerobic conditions:** It can be maintained by:

    i. Replacing the O_2 by CO_2 carbonation

    ii. Evacuating the sealed container (fruit juice)

    iii. Use of oils from top of the food (pickles)

    4. **Exposing at high temperature:** Fruits can be exposed by high temperature such as:

    i. **Canning:** Food is exposing to a high temperature (> 100°C) which prevents spoilage and inactivate the enzyme present in the food.

    ii. **Irradiation:** In case of irradiation, the food is exposed to the radiations to kill the survive m.o. by ionizing and non-ionizing radiation like a, b and g rays. Her, food is exposed to electromagnetic or ionizing radiation or various frequencies ranging from low frequency electromagnate to high frequency *i.e.* gamma rays which destroy the micro-organism present in the food.

B. Prevention/delay the shelf decomposition

i. By destruction or inactivate the enzyme – Blanching.

ii. Prevention / delay the non-enzymatic chemical reactions – Antioxident

Blanching

1. It is primary treatments which have to soften the tissues to facilitate packaging.
2. To preserve the original colour and flavour
3. To destroy the certain enzyme which are undesirable
4. Elimination of the air
5. Mostly for vegetables
6. Remove micro-organisms
7. Remove astringent taste and toxins

Anti-oxidant: Anti-oxidant are substances which are used to protect the food gamma deterioration caused by exposure to the air.

1. BHA – Butylactic Hydroxy Anisole Vegetable oils, BHT – Butylactic Hydroxy Toluene
2. Gellales: Animal fat, Vegetable oil
3. Tocopherols: Animal fat
4. Ascorbic acid: Fruit juices, Citrus oil, Wine, Bears *etc.*
5. Lactic acid : Processed fruits and vegetables, Canned fruits,
6. Phosphoric acid: Vegetable oils, Animal fat and cola drinks

Methods of preservation of fruits:

There are two main basic methods:

A. Bacteriostatic methods

B. Bactericidal methods

A. Bacteriostatic methods

1. Drying of foods
2. Use of chemical preservatives
3. Use of food additive
4. Use of low temperature

B. Bactericidal methods

1. Pasteurization
2. Cooking

3. Canning
4. Irradiation

A. Bacteriostatic Methods

In this method, the environmental conditions are change to prevent the growth of micro organisms, such conditions are called bacteriostic. These are -

1. Drying of Foods

Drying is just removal of moisture from the food to a certain level at which micro organisms cannot grow is called drying, it can be done by two methods:

i) Application of heat

a. Sun drying
b. Mechanical drying
c. Vacuum drying
d. Freeze drying

ii) Binding the moisture in the food

a. Use of sugar
b. Use of salt

i. Application of heat

a. **Sun drying:** Sun drying is the method in which food is directly exposed to sunlight. It is generally done in the places where plenty sunshine is available for long period e.g. Rajasthan. The dried product in this method is inferior in quality.

b. **Mechanical drying:** This is a method of drying where application of heat is applied by a mechanical dryer under the controlled conditions of temperature, humidity and air flow.

c. **Vacuum drying:** the temperature of the food and the rate of water removal are controlled by regulating the degree of vacuum and intensity of heat input.

d. **Freeze drying:** In this method, the food is dried by sublimation process, *i.e.,* just converting the food into ice without passing through the liquid form of water by means of vacuum plus heat applied in the drying chamber. In this method, product first frozen then water is removed by vacuum and application of heat which occurs simultaneously in same chamber.

ii. Binding the moisture

a. **Use of sugar:** The use of high concentration of sugar bind-up the moisture and make the food have a certain level of moisture at which micro organisms are not able to grow.

b. **Use of salt:** The concentration of salt causes the high osmotic pressure and tie up the moisture which inhibit the growth of micro organisms. It dehydrates the food by drying out and tie up moisture as it dehydrates the micro organism's cells. Salt reduces the solubility of O_2 in the food by reducing the moisture. It interferes with the action of proteolytic enzyme. The effectness of NaCl is varied with the concentration of salt and temperature.

2. Use of Chemical Preservative

Chemical preservatives are substances which are added to food just to retard, inhibit or arrest the activity of micro organisms such as fermentation, pacification and decomposition of the food. Chemical preservatives are of two types:

Class-1 preservatives: Common salt, sugar, dextrose, spices, vinegar, ascorbic acid

Class-2 preservatives: Benzoic acid and its salt, SO_2 and the salts of sulphuric acid, nitrates, sorbic acid and its salts, propeonic acid and its salts, lactic acid and its salts.

Among the class-2 preservatives, only two chemical preservatives are used in fruits and vegetables Preservation:

i. KMS

1. It releases the SO_2 and it is unstable.
2. It is used for the fruit which have non water solvent pigment (colourless).
3. It cannot be used in naturally coloured juices such as phalsa, jamun because they have the Anthocynin pigment.
4. It cannot be used in the product which are packed in container because it acts on the tin containers and oil, Hydrogen Sulphide (H_2S) which has an unpleasant smell and also form a black compound with the base plate of containers.
5. Best to control moulds than bacteria.
6. 350 ppm KMS is mostly used in fruit juice products.

ii. Sodium Benzoate

1. It is salt of benzoic acid and soluble in water.
2. It delays the fermentation in the juices.

3. It is commonly used in the product which are having natural colour such as anthocynin pigment.
4. It is more effective against the yeast.
5. 750 ppm Sodium benzoate is mostly used in fruit juices, squashes and cordials.

3. Use of Food Additive

Food additives are substances or mixture of substances other than basic foodstuffs, which are present in the foods as reagent of any aspects of production, processing, storage, packaging *etc.* Food additives are (i) sugar, (ii) salt, (iii) acids, (iv) spices. In case of sugar and salts, they exerts osmotic pressure by water is diffuses from the product through a semi-permeable membrane until the concentration reached equilibrium. They kills the micro organisms or do not allow them to multiplication.

i. **Sugar:** The concentration of 68-70% is used for preparation of jam, jelly, marmalades *etc.* sugar act as a preservative by osmosis and not as a true poison for micro organisms. It absorbs most of the available water, so little water available for the growth of micro organisms.

ii. **Salt:** the concentration of salt 15-20% is used for the preparation such as pickles. Salt inhibits enzymatic browning and discolouration and also acts as an anti-oxidant. It exerts its preservative action by:

    1. Causing high osmotic pressure resulting in the plasmolysis of microbial cells.
    2. Dehydrating food and micro organisms by tying up the moisture.
    3. Ionizing to yield the chloride ion which is harmful to micro organisms, and
    4. Reducing the solubility of oxygen in water, sensitizing the cells against CO_2.

iii. Acids

1. Many processed foods and a beverage needs the addition of acids to impart their characteristic flavour and taste in the final product because acids provide desired flavour and taste.
2. They adjust the sugar and acid ratio in the food.
3. Proper balance flavour of the food.
4. They are also playing the role for controlling the pectin-gel formation.

Main acids are the following

1. Acetic acid (Vinegar)
2. Citric acid (Lime juice)
3. Lactic acid (Lactose)

1. **Acetic acid:** it is commonly used for pickles, chutney, sauce and ketchup, just to inhibit the growth of micro organisms.
2. **Citric acid:** It is used for preparation of jam, jelly, squash, nectar *etc.* just to increase the acidity.
3. **Lactic acid:** It is used for the formation of curd from milk, raw flavour, and specific to pickles.

iv. Spices

1. Spices are plant products which are used in flavouring the foods and beverages to enhance the food flavour, colour and palatability.
2. They acts as antibacterial and antifungal activity.
3. They impart as colour agent.

4. Use of Low Temperature

Low temperature retards the microbial growth and enzyme reaction because it retards the chemical reactions. This is not a permanent method because some micro organisms can also grow at low temperature.

1. Cellar storage: (Above 15°C)
2. Refrigerated storage: (0 to 5°C)
3. Freezing storage: (-18 to -40°C)

1. **Cellar storage:** These are the underground room where surplus food can be stored for sometimes, only root crops such as potato, onion can be stored for a limited period.
2. **Refrigeration:** Fruits and vegetables can be stored for 2-7 days. Semi-perishable crops, such as potatoes, apples *etc.* can be stored, in the commercial cold storage with proper ventilation, automatic controlled temperature for one year.
3. **Freezing:** It ties up the moisture and increase the concentration of dissolved substances in the food. But, sometimes enzymes are active even below the 0^0C. In this case before freezing, 'Blanching' is necessary for vegetable freezing.

B. Bactericidal Method

In this method, food material is exposed to higher temperature and high temperature helps to killing of the micro organisms due to coagulation of protein. It helps in inactivation of enzyme. Here moist heat is more effective than dry heat. At low pH high temperature is required than the high pH. High temperature can be employed by following methods:

i. Pasteurization: Below 100°C
ii. Boiling/ Cooking: at 100°C
iii. Canning: Above 100°C

i. Pasteurization

There are three methods of pasteurization

a. Bottle or holding pasteurization

This method is commonly used for the preservation of fruit juices at home. The extracted juice is strained and filled in bottles, leaving sufficient head space for the expansion of the juice during heating. The bottles are then sealed air-tight and pasteurized.

b. Overflow method

Juice is heated to temperature of about 2.5^{0}C higher than the pasteurization temperature and then filled in hot sterilized bottles up to the brim, during filling and sealing the temperature of juice should not fall below the pasteurization temperature.

c. Flash pasteurization

The juice is heated rapidly to a temperature of about 5.5^{0}C higher than the pasteurization temperature and kept at this temperature for about a minute. This method commonly used for canning of natural orange juice, grape and apple juices. It is a mild heat treatment, by pasteurization milk is pasteurized by HTST at 72^{0}C for 15 sec. Fruit juices are pasteurized at such temperature and for such periods as would render them sterile, without impairing their flavour. Usually, the juices are pasteurized at about 85^{0}C for 25-30 min., according to the nature of the juice and the size of container. Acid fruit juices require lower temperature and less time for pasteurization than the less acid ones.

Juices can be pasteurized in two ways

1. By heating the juice at a low temperature for a High time (LTHT) and
2. By heating the juice at high temperature for a short time only (HTST).

ii. Boiling/cooking

The primary objective of cooking is to produce a palatable food. Cooking results in:

1. Destruction or reduction of micro organisms and inactivation of undesirable enzymes.

2. Destruction of potential hazard in the foods which are present naturally through micro-organism.
3. Improvement of colour, flavour and texture of the food.
4. It improves the digestibility of food component.
5. Putting the temperature about 100^{0}C. by this method, food can be preserved for 10-24 hours at low temperature.

iii. Canning

Canning is done at or above 100^{0}C. In case of fruits which are acidic, they are canned at 100^{0}C, while in case of vegetable those are non acidic, they are canned at above 100^{0}C. Here, high temperature can be obtained by using steam pressure, time is varying according to the type of foods. Due to anaerobic condition any survivable organism will not grow.

On the basis of acid, foods are divided into four different groups

1. Low Acid Foods 5.3 and above Peas, Corn, Lime beans, Meat, fish, Poultry and Milk.
2. Medium Acid Foods 5.3-4.5 Spinach, Asparagus, Beets and Pumpkin.
3. Acid Foods 4.5-3.7 Tomatoes, Pears and Pineapple, Sauce.
4. High Acid Foods Below 3.7 Berries and Sauer kraut, Pickle.

Section II: Fruit Production Technology

18

Mango

Botanical Name	: *Mangifera indica*
Family	: Anacardiaceae
Chromosome No.	: 2n = 40
Type of Fruit	: Drupe
Edible Portion	: Mesocarp
Origin Place	: Indo Burma Region (Myanmar)

Other fruits related to Anacardiaceae Family

Cashew Nut	: *Anacardium occidentale*
Pistachio Nut	: *Pistacia vera*
Chiraungi	: *Spondias cytherea*
Aamla	: *Spondias pinnata*

Origin and history: Mango is the king of fruits which is grown in India for over 400 years. More than 1000 varieties exist today. It grown in almost all state, India's share is about 56% of total mango production in the world. Its production has been increasing since independent contributing 39.5% of total fruit production in India. Andhra Pradesh top in the production and Uttar Pradesh is top in area. Andhra Pradesh, Bihar, Uttar Pradesh, Karnataka, West Bengal and Gujarat together contribute for about 82% of total production.

Climate and Soil: Mango thrives well in tropical and subtropical climate. High humidity, rainfall and frost during the flowering period is harmful. The favourable temperature for its cultivation is about 24-27^{0}C however, it can tolerant to the temperature up to 48^{0}C.

Mango can be grown in wide range of soil however deep, well drained soil is most favourable for its cultivation. Alkaline soil, rocky soil and water logged soil should be avoided. The favourable pH of soil is 5.5-7.5.

Important varieties: More than 1000 varities in mango are existing today but for facilitating the study the varities are classified as follows.

Varieties suitable for North India: Dashehari, Langra, Chausa, Bombay Green etc.

Varieties suitable for South India: Banglora, Neelam, Swarn Rekha, Parie, Banganpalli, Mulgoa, Alphonso etc.

Varieties suitable for East India: Himsagar, Zardalu, Krishnabhog, Gulabkhas etc.

Varieties suitable for West India: Alphonso, Parie, Keshar, Malkurad, Jamadar etc.

Some important hybrid varieties

Mallika = Neelam x Dashehari (IARI New Delhi)

Amrapali = Dashehari x Neelam (IARI New Delhi)

Arka Punit = Alponso x Banganpalli (IIHR Banglaur)

Arka Aruna = Banganpalli x Alphonso (IIHR Banglaur)

Ratna = Neelam x Alphonso (F.R.S. Banglore, Andhra Pradesh)

Sindhu = Ratna x Alphonso (Marashbara Krishi Vidyapeeth, Konkan)

Prabha Shankar = Bombay green x Kalapaddi (A.R.I. Sabaur, Bihar)

Mahmood Bahar = Bombay green x Kalapaddi (A.R.I. Sabaur, Bihar)

Jawahar = Gulabkhas x Mahmood Bahar (A.R.I. Sabaur Bihar)

Mangira = Rumani x Neelam (F.R.S. Singareddy Andhra Pradesh)

Arka Anmol = Alphonso x Janardan Pasand (IIHR Banglaur)

Pusa Peetamber = Amrapali x Lal Sundari (IARI New Delhi)

Pusa Arunima = Amrapali x Sensation (IARI New Delhi)

Pusa Lalima = Amrapali x Sensation (IARI New Delhi)

Pusa Pratibha = Amrapali x Sensation (IARI New Delhi)

Pusa Shreshtha = Amrapali x Sensation (IARI New Delhi)

Pusa Surya = Selection from the variety Eldon (IARI New Delhi)

Ambika = Amrpali x Janardhan Pasand (IARI New Delhi)

Sai Sugandha = Totapuri x Kesar (MPKV, Rahuri, Maharashtra)

Some polyembryonic varieties of mango: Bappakai, Chandrakaran, EC-959862, Kensington, Kitchener, Kurukkan, Olour, Nakkare, Mynepinian, Mudan, Vellaikulumba, Starch, Pryor, Peach, etc.

Propagation: Mango can be propagated by both sexual and asexual method. In asexual method inarching, approach grafting and veneer grafting (June-Sep/Oct) is more common but cutting, budding and layering also possible. In mango propagation certain dwarfing rootstock like Totapuri, Taimuriya, Olour, Rumani,

Nakkare, Kurukkan are commercially used and vigorous rootstock like Dashehari, Chausa, S.T. 9 and Muvandan are commercially use.

Planting: Planting should always being done during monsoon period, but if irrigation facility is available planting can be done in the month of February-March. Normally pit size is 1 x 1 x 1 m and distance between plant to plant and row to row is 10 m but in case of Mallika and Amrapali distance will be 3 x 3 m and in case of Arka Aruna distance will be 5 x 5 m (row x plant).

Digging and filling of pit: For mango panting pit should be dig in the month of April-May and after soil solarization dig are filled with the help of soil compost, neem cake and some soil born pesticide like phorate, themate, furadon, carbofuron etc. at the time of digging of pit the upper and lower soil shoud be keep separetaly. At the time of filling of pit the upper soil filled first then lower soil. The level of field soil should be high up to 4-5 inch. After irrigation soil will be settal down and level of soil surface can maintained.

Manure and fertilizers: For one year plant 10 kg. FYM, 100 gm. Nitrogen, 70 gm. Phosphorus, 100 gm. Potash should be apply. Bearing trees may be given 750g N, 200g, P_2O_5 and 1150g K_2O/year/tree. It is better always to apply organic manures during October. Manures should be applied in a small trench dug from about 1.5-2m from the trunk up to the drip line. Watering should be done soon if no rains. Fertilizer application should always being done in the outer periphery of canopy because the feeding roots of the plant are there.

Irrigation: Irrigation should be according to the soil and weather conditions. Bearing trees should be irrigated regularly at 10-15 days interval from fruit set to maturity. Plant should be given rest by stopping irrigations at least 2-3 months before flowering for maximum fruit bud development. Under drip, plants may be applied with 40 liters/tree twice a week.

Pruning and training: Mango needs no regular pruning except removing dead and diseased branches. Young plants should be trained properly to have a good framework.

Flowering and fruiting: In mango flower panicle is mostly from terminal shoot and rarely from the auxiliary bud. Flowering time of mango is closely associated with the time of flower bud differentiation that, vary from the cultivar, climatic condition, management practices etc. In most of the cases flower bud differentiation occur during October- December, but in case of barahmasi varieties it occur during May-June and some time September-October. Flowering season therefore extend from November- February. On a mango tree early flowering may be seen on eastern and southern side. Having a higher sex ratio (that is lower number of perfect flower in comparison to male flower) late emerging panicle have a higher percentage of flower.

The proportion of perfect flower and male flower on a panicle is influence by many factors.

- The cool environment during the period of panicle development caused higher percentage of male flower and the warmer environment have higher percentage of perfect flower.
- The panicles located in the inner canopy of the tree have higher proportion of perfect flower than located on the periphery.
- The higher proportion of perfect flower is usually in the terminal pore of the panicle.
- The young mango tree may have lower portion of perfect flower than the adult tree.

Mango is a cross pollinated crop, flower are andro-monoecious, protogynous, entomophilous and are pollinated by common housefly. The number of flower per panicle varies between 1000-6000 it's depend upon the varieties. [Dashehari, Langra, Chausa and Bombay Green are self incompatible type (Sporophytic type)].

Physiological disorder of mango

1. Alternate bearing

The tendency of bearing habit of in one year followed by a very little or no crop in succeeding year is termed as biennial or alternate behaviour of the crops. It is due to genetically/varietal factors, growth pattern, C:N ratio, hormonal factors, climatic etc.

2. Black tip of mango

This disorder is more common in Uttar Pradesh, Punjab, Bihar and West Bengal. Distal end of the fruit term black and spot get harden. This disorder is mainly due to Boron deficiency. The black tip disorder is generally appearing in those orchards which are located in visibility of brick kiln. The fume of kiln along with CO_2 and SO_2 this could cause the problem.

Control

- Spray of Borax 0.6% before flowering and two more spray at gap of 15 days.
- Spray of washing soda 0.5% solution and castic soda 0.8% solution are equally good in minimizing the disorder.
- Allow brick kiln at a distance of at least 1.6 km. in the east and west and 0.8 km. in the north and south of mango orchard.
- Increase the chimani height 15-18 meter.

3. Biennial bearing in mango

Mango producing good crop one year and no crop or fewer crops in the next year is known as biennial bearing or alternate bearing. This is genetic and inherent in mango varieties. Major causes for this are supposed to be:

1. Climatological factors: Rain, high humidity, low temperature making on to off year
2. Age and size of shoots: Shoots of 8-10 months maturity will be productive.
3. Carbon/Nitrogen ratio: High carbon/moderate N encourages flower bud formation (30-40).
4. Hormonal balance: Higher levels of auxin and inhibitor like substance and lower levels of gibberellins like substances were found to be vital for a flowering shoot.

Inspite of several studies, the biennial bearing is still an unsolved problem which is thought could be corrected by genetic engineering only.

Presently the suggested measures to overcome biennial bearing are:

1. Proper upkeep and maintenance of orchards.
2. Deblossoming in on year (NAA).
3. Smudging and chemical regulation like application of paclobutrazol (10 gm/ tree), spraying 1-2% KNO_3, 6-8% calcium nitrate, etc.
4. Pruning: Pruning the fruited shoots and opening tree top properly.
5. Growing regular bearing cultivars: Bangalore, Rumani, Neelum and almost all hybrids.

4. Leaf scorching

This is due to potassium efficiency or chloride toxicity.

Control

- Collect and destroy the fallen leaf.
- Apply the potassium sulphate.
- Avoid planting of mango in saline soil.

5. Spongy tissue

This melody has been reported only in Alphanso. This is arising disorder accrued by inhibition of ripening. The certain portion of the pulp caused by in activation of ripening enzyme. This is due to high temperature, post harvest exposure of sun light etc. Spongy tissue inhibit the lower activities of amylase glutamate, dihydrozine, peroxidise, catalyse, and much higher activity inverts and ascorbic acid oxidase as compare to healthy tissue.

Control

- Fruit should be harvested as 2/3 maturity stage.
- Fruit should be avoided to exposure from light.
- Growing spongy tissue free cultivar.

6. Cluster in mango

It is locally called Jhumka. This melody is characterized a cluster of fruit. This is due to adverse climatic condition during the month of February-March. Where embryos are shrivelled and could not develop further.

7. Soft nose

Soft nose are caused by Ca deficiency, causing breakdown of flesh towards the apex of the fruit before ripening.

8. Mango malformation

Mango malformation first discovered by G. Watt 1891 in Darbhanga district of Bihar on the basis of symptoms there are two type of malformation.

i. **Vegetative malformation-** Which is most common in nursery stage resembles "bunchy top" which may dry and die in due course. The main symptom are found in three way that is : 1. Complex leaf bunch 2. Crowding of small leafless 3. Stunted plant growth.

ii. **Floral malformation:** In this malformation the more number of male flower and the main symptom are found in three way that is : 1. Large panicle 2. Thick peduncle 3. Less percentage of perfect flowers.

Causes of malformation

- It's may be fungal (*Fusarium monilifarmae*).
- It's may be viral.
- It's an acarological problem.
- It's may be due to malfarmean like substance (malfarmean- is a plant growth inhibitor).
- It's due to mangiferien like chemical (It is a phenolic compound).
- It's may be bio-chemical or hormonical.

Control

- Application NAA 200 PPM.
- Spray of fungicide like bavistin @ 0.2%.
- Spray of acaracide, rogar, die-methylene or phosphamidon or dimethoate, diamicron @0.2%.
- Pruning of effected part.
- The plant material should always been taken from free mother plant.

9. Fruit drop: Fruit drop is natural and is very high in mango especially during the first four weeks.

1. Soon after flower opening
2. After pollination and fertilization
3. At grain stage of the fruit.

This occurs as an adjustment to the resources available in the plant for the development of fruits and is natural. Drop of grownup fruits is a major problem. This may be due to competition between developing fruits, drought or lack of irrigation, adverse weather conditions and incidence of serious pests and diseases. This can be avoided by regular irrigations during fruit development, application of optimum doses of nutrients effective control of pests and diseases and some hormonal sprays like 2,4-D (10-30 ppm) NAA(5-50 ppm), 2,4,5-T(20 ppm) etc.

Insects and pest

Mango hopper *(Amritoides atkinsoni):* Insect sucks the sap from the tender part and its secreat sweet sticky substances on which shootymould developed. It can be control by spray of carbolic @ 0.5% or monocrotophas (nucron) @ 0.4% or phosphamidon @ 0.5%. The spray should be given after the emergence of panicle but before flower opening and second spray should be done when fruit attain pea size and third spray should be done 15 days after second spray.

Mango mealy bug *(Drosicha mangiferae):* Mealy bug suck the sap from the tender part during February-March. They affect the crop during flowering. Fertilized female crown down from tree during the month of April-May and lay egg in the soil around the tree trunk. Up to December-January and climb up and settle down. On the young shoot, twig, flower and young fruit, they suck the shape and effect plant part dry up and also cause in mature fruit drop.

Control

- Dig the soil around the mango trunk during hot summer. Clean the weed. After monsoon mix the methyl parathion dust @ 250 gm/tree.

- Do banding of trunk in the month of November- December with slippery band of alkathine sheet or sticky band of grease.
- Spray carborile 0.2% or monocrotophos 0.4%.

Mango Scale *(Aspidiotus destructor):* The mango scale sometimes appears as a pest in certain locality by sucking the sap from leaves. It can be control by spray of metacid @ 125 ml/litter of water.

Fruit fly *(Dacus dorsalis):* Maggot feed on pulp and converts it in to off smailing rotten semi liquid. The affected fruit begin to rot and fall down from trees. The adult fly emerge in April from cocoon in the soil and after mating the lay eggs in cluster 150-200 under the skin in the fruit just before ripening. After 2-3 days the maggot hatch out and start feeding on the pulp. The life cycle is completed in about 10-27 days during summer.

It can be control by collect and destroy the fallen fruits and spray of carbaryl 0.2% + 0.1% protienhydrolysate. The spray may be repeated after 21 days.

Stem borer *(Batocera rufamaculatd):* Mango stem borer is a serious past in mango. The grubs feed inside the stem from down to upward. Adult beetle are grayish brown in colour with brown and black dark. The adult beetles emerge during July- August. It can be control by application of petrol or kerosine oil.

Bark eating caterpillar *(Indarbala quadrinotatd)'.* The larvae of this insect are principle feeder of the bark and spul a loose brownish silken strands in which excretal pellets and piece of bark are entrapped. It has single generation in a year. The moths are observed during April-May larvae develop from May-September. Pupation occurs during February-March. The attack is more serious on old tree and neglected orchard. It can be control by application of petrol or Kerosine oil in to the whole during September-October and again in January-February.

Disease

Powdery mildew *(Ordium mangiferae):* It is the most serious disease in all most all part of the country. The disease has been reported to cause heavy loss of crop in the year, when humidity is very high a companied by cool night bearing the flowering. During flowering season in the beginning appearance of grayish whitish powder on the flower bud, frontlets and rachis of the panicle, which latter become dark brown in colour. It can be control by Sulphur fungicide. Application of Sulphex (wettable sulphur) 0.2% or bavistine 0.1% or karathene 0.1%. Sulphur is given just before panicle emergence and application of bavistine and karathene 15 days after.

Anthrecnose *(Colletotrichum gleosesproides):* It affects the tender shoot and foliage. Which result in the death of the affected portion. The symptoms vary according to the plant part affected i.e. dark brown necrotic area of the leaves, elongated black necrotic patch on twig, small dark spot on the main and secondary rachis of the panicle and the black spot of older fruit. It can be control by application of Bordeaux mixture (Copper sulphate: CaCoa: Water) 3:3:50 or copper oxyxhloride or dithiocarbamate 0.2% or bavistene 0.1% in the nursery.

Bacterial canker *(Xanthromonas compestris pv. mangiferae):* The bacteria blight disease has become wide spread on all most all varieties of mango. It assumes serious form only in humid area or in season when temperature and humidity conditions become favourable for its spreads from the leaves on to the fruits in early stage of development. Leaves as well as fruits both are damage but fruits become severely attack and disfigured on leaves, panicle and fruits tinny spot appear in the initial stage, these spot gradually in large in areas and later form a raised spot on one side on the fruit. It can be control by application of Streptomycin sulphate 200 ppm.

Harvesting and yield: Stage of harvesting is very important, indicated by

1. Starting of colour development
2. Falling of one or two fruits from the plant
3. Specific gravity of 1.0 to 1.02 (more reliable)

Mango normally takes 90-120 days from fruit set to maturity. Harvesting is done using pole harvesters without causing any damage to the fruit. Mango grafts come to bearing in about 2-3 years but commercial yields can be had from 8-10 years and may continue up to 40-60 years. Average yield is 8 tones/ ha and may vary according to variety and locality. On an average a full grown mango tree yielded 200-600 kg fruits per tree per year.

19

Grape

Botanical Name	:	*Vitis vinifera*
Family	:	Vitaceae
Chromosome No.	:	2n = 38
Type of Fruit	:	berry
Edible Portion	:	Pericarp and placenta
Origin Place	:	In between Black Sea and Caspian Sea

Introduction: Grape is an important sub-tropical fruit crop in India. The plant of Vitaceae family divided into two important groups first is Euvitis which have 60 species; some important species which are commercially grown are i.e. *Vitis vinifera, Vitis raperia, Vitis rupestris, Vitis labrausca* and *Vitis champini.* Second group is Muscadiana and in this group the important crops are *Muscadiana rotendifolia, Muscadiana popenoii* and *Muscadiana munsoniana.* The chromosome number of this group is 2n = 40.

Characters of Vitis and Muscadinia

Characters	***Vitis***	***Muscadinia***
Shoots	Bark is longitudinally striate, fibrose	Tight bark, non-shedding, with prominent lenticels
Pith	Interrupted in nodes by a diaphram	Without diaphram
Tendril	Forked	Simple
Flower clusters	Elongated	Short, small
Berries	Adheringtothe cluster at maturity	Detach one by one they mature
Seeds	Pyriform with long or short beak	Oblong without beak

The four important species which are originated in India are-

Vitis barbata – Assam, Bengal, Myanmar reason.

Vitis rugosa – Kumaoon hills

Vitis rumeisperma – Kasmir

Vitis purviflora – Sikkim

Origin and history: Grapes are commercially grown in America, France, Canada, Italy, Spain, and Pakistan. In India leading growing states are Karnataka, Maharashtra, Tamil Nadu and other states like Punjab, Uttar Pradesh, Himanchal Pradesh, Rajasthan, Madhya Pradesh etc are equally important.

Soil: Grape requires good loamy soil with good drainase facility. The light sandy to sandy loam soil is excellant for grape cultivation. During the active growth of the plant excess moisture is injurious for root growth. The soil pH should be range from 6.5-7.5. In general the grapes is succeptible to slanity and alkalinity.

Climate: Grape is a subtropical fruit crop and it is require specific climatic condition for better growth and development of fruits and their quality.

1. At the time of berry ripening atmosphere should not be dry other wise cracking of berry take place and TSS contain will be deteriorated.
2. For good rate of growth and development of the fruits and their internal constituent fruits require moderately cool to warm weather. If temperature is risese very Quickly it is not favourable. The slow rising temperature favours the development of chemical constituent of the fruit responsible for quality table grape.
3. Grapes required long warm and hot dry summer and cold winter. It dose not required humid summer because in humid summer there is a big chance of attack of insect pest and diseases (powdery mildew and downey mildew).
4. Area having early spring and late spring should be avoided.

Uses

Grapes are used for different purposes.

1. Table purpose

Table grapes are meant for use as fresh fruit consumption. These grapes are attractive in appearance and eating quality and with good shipping and keeping qualities. Most of the varieties grown in India are table fruits. The important table grape varieties are Muscat Humburg, Cardinal, Perlette, Thompson Seedless (Sultania), Tokay, Concord, Anab-e-Shahi, Pusa Seedless, Delware, Catawba, Ohanez, Red Malaga, Emperor, Italia, Muscat of Alexandria, etc.

2. Raisin making

These are the grapes intended for making dried grapes. The raisin variety of grapes should have soft texture, selflessness with good sugar content, marked pleasing flavour, large or very small size and little tendency to become sticky in storage. The varieties most extensively used in the commercial production of raisin include Thompson Seedless (Sultania, Oal Kishmish), Seedless Sultana, Red Corianth, Cape Currant and Black Monukha.

3. Juice making

The varieties of sweet juice grapes produce juice of acceptable beverage. The juice should retain the natural fresh grape flavour throughout clarification and preservation. In United States of America the Concord grapes are in general for Juice. The varieties White Riesling and Chasslas Dore are used for juice in the central Europe. The varieties Aramonand Carignan are utilized for sweet juice in France.

4. Wine making

Most of the vineyards in Europe, North Africa, South Africa, and South America, Australia and United States of America produce wine grapes. Wines are classified as table wines and dessert wines. Table wines contain less than 14 per cent alcohol while the desert wines have more than 14 per cent alcohol, usually 17 to 20 per cent sugar acid ratio, total acidity and tannin content etc., will determine the wine quality. The varieties such as White Rieslin, Chardonnay, Cabernet Sauvignon, Tinta Maderi and Muscat Blanc (Muscat Canelli) produce wines of high quality, outstanding in bouquet, flavour and general balance.

5. Canning grapes

Seedless varieties like Thompson Seedless and Canner are generally canned in combination with other fruits. The varieties grown in Tamil Nadu belong to table grapes. Pachadraksha, Muscat ("Panneer"), Anab-e-Shahi & Bangalore Blue are the main varieties.

Planting: Pit size 90 x 90 x 90 cm and filled with mixture of soil and manure. One year old plants are planted pruning the stem to 2-3 buds. Spacing varies according to the method of training and variety from 1.2 to 5.0 m X 2.5 to 6.6m. Pandal system requires maximum spacing. As the plant grows staking should be provided, best time for planting is Jan-Feb for rooted cuttings and October for un rooted *in-situ* planting. Planting distance 3 x 3 m, 4 x 3 m, 5 x 3 m, 6 x 3 m are depend upon variety, climatic condition, soil type and training and pruning system.

Manure and fertilizers: For one year old plant 10 kg FYM, 10 gm N, 60 gm P_2O_5 and 80 gm K_2O per year per plant. This dose should be increases every year in same amount upto 5 year of age then it should be stable. This full dose of phosphorus, half dose of potassium and nitrogen should be given just after pruning and the rest dose of potassium and nitrogen should be applied just after fruit set.

Propagation: The grapes are commercial propagated by two methods i.e. sexual and asexual. In asexual method the hard wood stem cutting is more common. The cuttings are always made from mature cane. The cutting should have at list 4 bud and 23-46 cm long. Besides cutting it can be propagated by mound layering, chip budding and grafting.

The important rootstock of grapes are:

Phylloxera resistant rootstocks are: Riparia Gloire- Selection from *V. riparia*, St. George - Cultivar of *V. rupestris,* 4x2 (Ganzian)- (*V.vinifera* (Aromon) x *V. rupestris* (Gonzin), 1202- (*V.rupestries* and *V. mataro*), 99- (*V. berlanderi* X *V. rupestris*).

Nematode resistant rootstocks are: Dogride- cultivars of *V. champin*, Salt creek, 1613- 1616- (*V. solonis* X *V. paria*), Telecki 5A- (*V. berlandarix* X *V. riparia* 5A 1616), Solonis 1616- suitable for saline & wet soils.

Salt resistant rootstock are: Salt creek

Important varieties: More than 20 varieties are under cultivation however, only a dozen of varieties are commercially grown. They can be grouped under 4 categories on the basis of colour and seed.

Coloured seeded variety: Bangalore Blue, Gulabi (Muscat).

Coloured seedless variety: Beauty Seedless, Sarad Seedless.

White seeded variety: Anab-e-Shahi, Dilkush.

White seedless variety: Perlatte, Pusa Seedless, Thomson Seedless, Tas-A-Ganesh, Sonaka, Manik Chaman. Currently Thomson Seedless is an important variety which occupies 55 % of total grape cultivation.

Important hybrid varieties

Arkavati = Black Champa x Thomson Seedless.

Arka Shyam = Bangalore Blue x Black Champa.

Arka Kanchan = Anab-e-Shahi x Queen of Wine Yard.

Arka Hans = Bangalore Blue x Anab-e-Shahi.

Arka Neelmani = Black Champa x Thomson Seedless.

Arka Shweta = Anab-e-Shahi x Thomson Seedless.

Arka Majestic = Angoor Kala x Black Champa.

Arka Chitra = Angoor Kala x Anabe-E-Shahi.

Arka Shoma = Anab-e-Shahi x Queen of Wine Yard.

Arka Trishna = Bangalore Blue x Convent Large Black.

Arka Krishna = Black Champa x Thomson Seedless.

Pusa Urvashi = Noor x Beauty Seedless.

Pusa Navrang = Madeleine Angevine x Rubi Red.

Quality regulation of grape: The quality of berry can be regulating by two methods i.e. mechanical method and chemical method.

Mechanical method: In this method ringing and girdling, berry thinning and bunch thinning are the important one. The basic objective of this method is to supply to increase the food material per berry or per bunch.

Chemical method: In chemical method, application of GA_3, Ethephon and NAA has been found most effective to increase the yield and quality of fruit. Among these, GA_3 30 PPM at full bloom stage has been found most effective.

Irrigation : In the following period irrigation should be avoided. If irrigation is needed it should only be maintain the low level of moisture.

i. From October-December, i.e. plant going toward the dormant condition only one irrigation is sufficient when soil is very dry.

ii. At the flowering time flower opening stage irrigation should be avoided because fruit setting will be adversely affected.

iii. At the time of fruit ripening the low level of moisture should be maintain in vineyard.

When irrigation is needed: Irrigation should be given just after pruning, after application of fertilizer and manure and at the time of fruit setting to fruit maturity.

Pruning and training of grape

Pruning: Pruning means removal of the plant parts without effecting the growth and development of the plant and quantity of the fruits etc.

Training: Training means giving a proper shape with the help of pruning is known as training.

Pruning

The grapevine is a vigorous climber. If it is not properly trained and pruned, it does not bear fruit properly. Pruning is one of the most important operations in grape culture. The objectives of pruning are as follows:

1. To reduce the amount of old wood in order to keep the vine within manageable limits.
2. To secure fruit bearing branches in predetermined places.
3. To expose the fruiting branches to sufficient sunshine.
4. To reduce the excessive vegetative growth.

It is essential that one should be familiar with the various parts of the vine and their functions for successful pruning. The following are some of the technical terms used:

Shoot: Young growth of green stem of the current season, which bears the grape cluster.

Cane: The young leafy growth of the current season in a shoot which after maturity is known as cane.

Spur: The basal portion of the cane which left after pruning is known as spur. It is cane pruned to 1-2 buds.

Fruiting spur: A cane or well ripened shoot leaving 304 buds, producing a bunch after pruning.

Foundation spur or renewal spur: A well ripened shoot or cane bearing bud. This normally remains after the shoots are pruned in March-April or summer in Hyderabad. It is called a **foundation spur** as it forms the base of the foundation wood on which next year's canes and fruiting spurs are formed or on which both growth of the year are borne.

Trunk: Main stem of the plant.

Long spur: A ripe shoot, carrying more than five buds. Normally it is 25-30 cm long with about 5-10 buds on it.

Medium spur: It is a cane cut back keeping 3-5 buds.

Fruiting cane: Spur intended to bear fruit normally with 2-4 bud that is fruiting spur and the basal portion of the mature cane with 8-12 bud that is left after pruning produce fruit is called fruiting cane.

Time of pruning: Pruning is done once in north India during the month of June while south India twice in a year once in summer (March-April) and again in winter (first week of October).

How much to prune: Some of the cultivar like Perlette, Beauty Seedless, Delight, Bangalore Blue and Early Muscat. Fruit arising from the basal bud of the cane, in such cultivar cane are heather back to 4-5 bud which is known as spur fruiting cultivar.

On the other hand Pusa Seedless, Thomson Seedless, Kismis Charny, Gulabi etc. cane is heather back to 8-12 bud these cultivar are called cane bearing cultivar.

Training

The training of the vine depends on two fundamental factors namely, the growth characteristics of a variety and the influence of the local climate on the growth of the variety. In addition, the training system selected should take cognizance of economic aspects also, such as the initial outlay on erection, subsequent maintenance and cost of production of the crop. Defective training may result in delayed bearing and irregular development of vines resulting in reduced yields. The reduction, in yield may also be due to the failure to utilize properly the full vigour of the vines. Training system adopted widely in India is the bower systems.

Ever since the introduction of grapes in Tamil Nadu, the grape has been trained in overhead arbours or pandals. The height of the bowers is however very low and the root effect of the stakes is an objection. Recently, growers are going in for stone and cement pillars or even G.I. tubes. The overhead canopy is made of either thin bamboos (2 to 2.5 cm) or G.I. wires of different thickness, (8 to 16 in guage). The cross piece connecting the pillars are made of thick bamboos or trek-wood or palmyra rafters or iron tubing. The wires are spaced (30 x 45 cm) or 30 x 30 cm apart forming a network.

The rooted cutting planted in the field reaches a height of 180 cm (6 feet) in about three months. When the main stem is pinched at the top that is about 15 cm below bower. Two to four side shoots are allowed to grow and are trained on either side of the trunk or on four sides to from the arms subsequently. These side shoots or main arms as they are called are allowed to reach the periphery of the pandal and they are then tipped. On these arms secondaries are produced and allowed to grow on the main arms at intervals of 45 cm from each other alternately in opposite directions. These branches are trained to cover the framework uniformly. The advantages of the pandal system of training are as follows:

1. The climate prevailing in India affords full scope for the vine to grow as luxuriantly as possible. Consequently, it facilitates harvests of a succession of crops at intervals of 4½-5 months.
2. The crop gets enough sun shine uniformly for their proper development.
3. Plant protection measures are more effective in this system, especially against mildew.

Training of grape: There are 4 training system applied in grape.

1. Head system
2. Telephone system
3. Pergola system
4. Kniffin system

The different training systems are described here under:

1. **Single-stake System:** A single shoot is allowed to develop from the vine of rooted cuttings and is trained vertically by staking to a support. When this shoot reaches a height of 120 cm. it is tipped and allowed to produce 4 to 5 secondary branches or canes, which are pruned after every bearing season. The main stem and the primary laterals are supported by a bamboo post planted nearby.

2. **The Four-arm Kniffin System:** In this method, the vine is allowed to put forth a single shoot which is trained erect and tipped at a height of about 45 cm from the ground level. Only three shoots are allowed to grow from this point, all others being removed. Two of the retained shoots are trained horizontally and the remaining is trained vertically to a height of another 60 cm. When it is tipped to a height of another 60 cm. When it is tipped again to produce two more shoots, which are trained horizontally. All the four horizontal arms are supported by bamboo poles or wire tied horizontally to posts fixed at regular intervals of 3 metres. Primary laterals are allowed to develop from these four main arms at fairly regular intervals and these are pruned to produce the crop every season. For every fruiting cane, a renewal spur of 2 to3 buds is left. Normally only four canes are allowed in the 4-cane kniffin system and 6 canes in the 6 cane system. Under tropical conditions it is possible to have doubled the number.
3. **Wire Trellis System:** In this system two or three wires are strung in rows from vertical posts. A single stem in between the posts is trained as far as the top of the wire 2.1 m. and two arms are allowed to develop along the wire on either side. Each arm will meet the arm from the adjacent vine and will have two arms with six tertilaries. The spurs on the fruit bearing shoots growing on these canes are seasonally pruned for fruit every year.
4. **Telephone Trellis System:** Because of its resemblance to the telephone pole with its flat topped mast bearing the supporting wires, this is called as Telephone Trellis system. In this system there is large flat-topped T-Trellis above ground level. The horizontal cross is 120 to 150 cm wide with wires strung on the top 30 to 45 cm apart. The vines are trained along the wire in the direction of the row. Steel angle iron or waste structural steel is used when available and the cross arm is welded to the upright. In a few vineyards the flat top is modified by bending each half of the mast upwards about 30° so that a trough shaped or 'H' shaped form is obtained. The disadvantage in this system is that there are no cross supports to the rows to withstand strong winds. Cultivation and movement is limited tone direction. But they are provided with better exposure to light.

Flowering and fruiting: In grape flower born in cluster which appears to be lateral to the shoot. Most of the *vinifera* cultivar have perfect flower but some species of grape ie. *Vitis rotundifolia* have dioecious flower. Berry set result from pollination and fertilization however; some of the cultivar develops fruit by parthenocarpy. In grape fruit develop by stenospermocarpy.

Physiological disorder: Due to in adequate leaf area and non availability of fruit reserve to developing cluster thinning, girdling and use of growth regulator can reduse the problem.

1. **Blossom:** Blossom a black suskan spot develops at the blossom end of the berry which later on spreads + water suncans and due to rot calcium deficiency.
2. **Interveinal chlorosis:** Mn, Zn or Fe deficiency-0.2%.
3. **Stalk necrosis:** Calcium deficiency.
4. **Bud, flower and berry drop:** Girdling 10 days before full bloom, 500 ppm ethrel at ripening NAA 100 ppm at 10 days before ripening, Reducing irrigation during bloom, Benzyl adenine 200 ppm, 4-CPA 20 ppm-for thinning.
5. **Bud killing:** Excessive nitrogen.
6. **Hen & Chicken:** Due to boron deficiency.
7. **Water berry:** Due to water stress or excessive of N_2 fertilizer or over bearing.

Post harvest fruit drop: This is due to weak pedicle attachment to the berry, this is more common in Anab-e-shahi, Beauty Seedless. The spray of NAA 50 ppm a week prior to harvest can minimize the problem.

Uneven ripening: This is due to in adequate leaf area and non availability of the food reserve to a developing bunch. The cultural practices like cluster thining, girdling and use of growth refgulators can be reduce uneven ripening. The applications of ethephone 250 ppm at colour break down stage recomonded to reduce the problem.

Pink berry formation: In this melody pink blush pink colour change to dull red colour and berry become soft and watery. Definite cause is not known and it is control by application of 0.2% of ascorbic acid and 0.25% solution of sodium diethyldithiocarbamate.

Insect and Pest

Scale: These are mostely found near the curve and near the branchase and suck the sap. It can be control by application of Monocrotophase 0.4%, pruning of seviourly affected branchase and old vine should be removed.

Chaffer beetle: It make a number of holes in the leaves and feeds the tendal leaves. It generally attacks in rainy season because this photosynthetic rate of the leaves is decreases. It can be controlled by Cypermethline 10 % or Carborile 0.3%.

Disease

Powdery mildew (*Unicinula necater*): In this problem leaves and fruits are mostely affected the white powder moss of the fungus appear on the upper portion of the leaves and fruits and letter on it tern in brown. It can be control by application of Bavistin 0.1% or Kerathane 0.1% at weekly interval.

Anthracnose (*Sphaceloma ampelina):* In this problem the brown spot appear which is large and in seviour casasse young shoots are dry and the affected leafs are drope down. It can be control by Dithane-Z-78 0.2% or Difolton 0.2% or Blitox – 50 0.3%.

Downey mildew (*Plasmospora viticola):* In this problem leaves, shoots and other portion are affected light yellow spot are form on the lower side. It can be control by spraying of Bordeux mixture 1% or Blitox 0.2%.

Harvesting and yield: Grape is a non climacteric fruit and has to be harvested at correct stage of maturity. Degree days from full bloom give a correct indication of maturity. Early cultivars require about 1600-2000 degree days and late cultivars about 3000 or more degree days. A cluster having under developed seedless brries known as "shot barriers" or mummies. These will be very sweet in taste. Average fruit yield is 30 ton/ha. (300 q), where on the cultiver Anab-e-Shahi – 97.04 ton/ha., Chima Sahebi – 91.39 ton/ha are found.

20

Ber

Botanical Name	:	*Zizyphus mauritiana* (Indian ber)
Faimly	:	Rhamnaceae
Chromosome No.	:	2n -24
Type of Fruit	:	Drupe
Edible Portion	:	Pericarp
Origin Place	:	Indo-China region

Origin and history : Ber is an ancient fruit crop of India and is relished for its sweet and sour fruits. It is an ideal fruit tree for arid and semi arid regions in tropical and subtropical climate. It is highly suitable for marginal land and hot arid region. It is distributed in tropical and sub tropical country in the world, including China, Pakistan, Australia and France. In India the major growing states are Madhya Pradesh, Bihar, Uttar Pradesh, Punjab, Haryana, Rajasthan, Gujarat, Maharashtra, Andhra Pradesh etc. In Uttar Pradesh, Varanasi, Aligarh, Ayodhya, and Agra are the major growing region. Beside *mauritiana* other important species like *Zizyphus rotundifolia*, *Zizyphus numuleria, Zizyphus rugosa, Zizyphus xylopyrus, Zizyphus micropyla* is also commercially grown for rootstock purpose.

Important characteristics of the ber which are suitable for arid region plant:

1. Plant should have deep tap root system.
2. Plant should have waxy coating on the arial part.
3. Plant should have capacity to bind water (Maximum water binding capacity is found in Fig).
4. Plant should have characteristic to shading of leaf at the time of scanty sunlight.
5. The flowering and fruiting of the plant are shrinconized with the maximum availability of moisture in the atmosphere.

Climate and soil : It prefers hot and dry climate for quality fruits and production. It can be grown up to an altitude of 1000 m from MSL. It can be grown in all type of climate but it is sensitive to freezing temperature. Because of its hardy and taproot system, it can be grown in a wide range of soil including porous and infertile soils unsuitable for major fruits and other crops. Soil is no limiting factor for ber cultivation it can be grown in all type of soil it can tolerate salinity alkalinity and water logging condition.

Species and varieties

The genus Zizyphus consists of 40 species in tropical and subtropical regions. *Zizyphus jujuba* (Chinese ber) and *Z. mauritiana* (Indian ber) are the two species grown widely in temperate, tropical and subtropical regions of the world respectively.

The important species of Zizyphus in India are as follows

- *Z. mauritiana, Z. jujube*
- *Z.nummulaia/Z. rotundifolia*
- *Z. oenoplia*
- *Z. rugosa*
- *Z. sativa*
- *Z. xylopyrus*

Important variety

Early season- Gola, Seb, Safeda, Sandur, Narnaul, Seo, Choncahl, Shamber, Badami, Manuki, Guli, Nazuk, Noki, Rohataki, Selected Safed, Sandhura, Narnaul, Narma Varanasi, Banarasi Gola, Delhi etc.

Mid season- Kaithli, Sanaur-5, Muria, Mehrara, Mehrun, Kharki, Darokhi, Banarasu, Dandan, Kaithii, Walaiti, Thorn Less, Jogia, Mundia.

Late season- Umran, Illichi, Katha, Jogia, Aliganj, Dandan, Chameli.

Propagation- Existence of old orchard in India is slightly difficult due to the hard endocarp and requires about three to four weeks to germinate. The germination time can be reduced to one week by cracking the endocarp carefully. The seedlings raised are mainly used as rootstocks. Ber is commercially propagated by ring budding and T budding best time of propagation is July – September.

Planting- A plant spacing of 6m x 6m is followed for ber and where on the digging pit is 1m x 1m x 1m. Planting should be done in monsoon period i.e. June-July (South India) and August-September (North India).

Manure and fertilizers- One year old plant require 10 kg FYM, 100g N, 50g P_2O, 50g K_2O, this dose is increasing up to the 5 year in the same fashion and 5 year old plant needed 50 kg FYM, 500 g N, 250 g P_2O and 250 g K_2O.

Irrigation: First irrigation is just after the transplanting then later irrigation is given as per requirement of the plant. In general in summer irrigation should be given 8-10 days interval.

Flowering and fruiting: Time and duration of flowering varies with cultivar and location. Flowering starts from July and extended upto November in different regions of the country. In south India, flowering starts from July and continues to September with major fruit set in August it self, flowering start in north India, in the month of September and up to November. Ber is a cross-pollinated crop and most of the varieties are self unfruitful that is why pollinizers is require for proper pollination and main pollinating agents are honey bee. Fruit growth of ber followed a double sigmoid curve with three distinct phases i.e. early phase of 45 days after fruit set (DAFS), middle phase, 45 to 90 DAFS and final phase, 90 DAFS and to maturity (185 days).

Insect and Pest

Fruit fly (*Carpomyia vesuviana*): This is the most dangerous insect of ber crop infestation starts in September when the fruits are very small and continue upto harvest. The adult female lays eggs singaly by inserting its ovipositor in the developing fruits. This insect control by the summer ploughing of the orchard, application of 0.1 % aldrin or 0.1 % Rogor 30 EC or 0.04 % monocrotophos.

Bark eating caterpillar (*Inderbela quadrinotata*): Bark eating caterpillar is an other important insect which is damage the bark of the trunck and branchese and their control is very cheap and easy. It can be control by kerosine oil or petrol.

Disease

Powdery mildew (*Oidium erysiphoides*): This disesese can be seen in the form of white powdery growth on flowers and fruits and rarely on leaves also it appear on fruits during October to November after the fruit set is over.

In initial stage it can be control by spray of 0.2% kerathene or any sulphur fungicide like sulfex 0.1%.

Harvesting and yield: To get quality fruits, fruits should be harvested at correct stage of maturity. Under ripe fruits do not ripen properly and never attain the desired quality. Over-ripe fruits, on other hand, lose their attractive colour and become soft. Such fruits have less storage life and deteriorate in quality. The fruit are ready to harvest in the month of Feb-March and the average yield is about 80-250kg per tree.

21

Guava

Botanical Name	: *Psidium guajava*
Family	: Myrtaceae
Chromosome No.	: 2n = 22
Type of Fruit	: Berry
Edible Portion	: Pericarp
Origin Place	: Tropical America

Origin and history: Guava (Amrud) is one of the most common fruits in India. It is a very hardy sub-tropical plant and prolific bearer. In India, it is commonly called as "poor man's apple" widely naturalized in the country. Allahabad has the reputation of growing the best guava in the country as well as in the world. Area wise guava comes under 4th position after Mango, Banana, Citrus, in India. Uttar Pradesh has the largest area fallowed by Bihar, Madhya Pradesh, Maharashtra, Gujarat, and West Bengal, however it is generally grown all over India and also grown in South Asia that is Srilanka, Myanmar and tropical and subtropical country. Beside *Psidium guajava* some other important species which are commercially grown for rootstock purpose that is :

1. *Psidium montanum*
2. *Psidium friedrichsthalianum*
3. *Psidium cattleianum*
4. *Psidium chinensis*
5. *Psidium polycarpum*

Guava used as fresh and also use in Jelly, Jam, Nector and Beverage, etc.

Composition and uses: The fresh fruits are very rich in vitamin C (100-260 mg/100g pulp). Vit-A, B_2, and minerals like Calcium, P, acidity 2.4 per cent, carbohydrates 9-10 per cent, TSS-13° Brix, pantothenic acid, riboflavin, thiamin and niacin, also rich source of pectin.

- The fresh ripe fruits are used as table/salad fruits.
- Fruits are rich in pectin. The best quality jelly can be prepared; fruits can be canned in sugar syrup or made in to fruit butter, juice preparation and in ice-creams.
- The leaves yield a dye and are used in dying industry and also have medicinal values for curing diarrhea.

Soil and climate: Guava is resistance to adverse soil condition and can be grown from pH up to 4.5-9.0 but for better cultivation pH should not be more than 8.5. However, it performs well in fertile soil. Guava tree fruits only in new growth. Well drained fertile loamy soil are cosiderd to the highly suitable.

It can be grown successfully in tropical and mild subtropical climate the optimum temperature is 15-30^0 C but it is sensitive to frost and can tolerate high temperature up to 46^{0}C.

Species: There are more than 150 species available in guava and some of the important species are

1. *Psidium guajava*- It is the commercially cultivated species, rest of them do produce fruits but small size, inferior quality and with high acid content.
2. *P. guineense*- Guinea guava – has small fruit with poor quality.
3. *P. guajava var. aromaticum*- small scented fruits.
4. *P. pomiferum*- fruits are round.
5. *P. pyriferum*- fruits are pear shaped.
6. *P. cattleianum*- the strawberry guava with round red fruits.
7. *P. friedrichsthalianum*- Chinese guava with small and globose fruits having high acid content and resistant to wilt.

Important variety: Allahabad Safeda, L-49 (Sardar guava-develop by Dr. G.S. Cheema), Red Flesh, Chittidar Guava, Apple Colour, Nashik, Dharwar, Arka Mridula, Allahabad Surkh, Varanasi Surkh, Arka Amulya, Hisar Safeda, Hisar Surkh, Pant Prabhat, Lalit, CISHG-1,2,3 and Sangam (Develop by Directorate of Horticulture U.P) etc. Hybrid variety are Safed jam- Allahabad safeda X Kohir, Kohir safeda- Kohir X Allahabad safeda (Fruit Research Station, Sangareddy, Andhra Pradesh).

Propagation: It can be propagated by both sexual and asexual method. In India and in many other countries guava is commonly propagated from seed which is quick and cheap method. Seed retain there viability for one year and germinate within 2-3 week. In asexual method patch budding, forkert budding, side veneer grafting, inarching, approach grafting and mound layering or stooling is commonaly practice.

Land preparation

Raising of seedlings: Rootstocks are raised from the seeds extracted from healthy fruits and stored for 100 days by treating with ferulic acid at 10^{-3} Molar concentration. Potassium nitrate also used at 1 per cent to prolong storage of seeds. The seedlings will be ready for grafting in 45-60 days after sowing, the layers should be treated with IBA 10,000 ppm for better rooting.

Planting: Pit size, 1m x 1m x 1m. Planting is done during monsoon at a spacing of 8m x 8m or sometime 6m X 6m provided with 30-40 kg FYM, dried leaves and 1 kg SSP. After planting they are supported by stakes. Appropriat time for planting is the onset of monsoon i.e July-August

Irrigation: In guava for high quality fruits mrig bahar is recomonded for this crop require 2-3 post monsoon irrigation during winter at an interval of 15-20 days. During summer and winter season orchard is irrigated at an interval of 4-6 and 10-15 days respectively. First irrigation should be applied just after transplanting.

Manure and fertilizers: For one year old plant 10 kg FYM, 60 gm N, 30 gm P_2O_5, and 60 gm K_2O per year. For 6 year old plant required 60kg FYM, 360 gm N, 180 gm P_2O and 360 gm K_2O. Zinc and boron difficiency also reported in guava. 0.5% Zinc Sulphate and 0.5% Borax are recomonded twice in a year first is June-July and second in October is beneficial. The application of urea upto 0.2-0.4% is found most effective.

Training and pruning: Training is done primarily to give frame to the tree. For development of a strong framework, the first 60 to 90 cm from base of the trunk should be cleaned followed by 4 to 5 scaffold branches at an interval of 20-25 cm. When the plants attained a height of about 1.5m to 1.8 m, it is headed back to make the center open. In some parts of India (Maharashtra and South Bengal), the branches are bent down ward and tied to each other thus forcing the dormant buds to grow. This results in increased yields.

The trees are rarely pruned in North India, but light annual pruning after harvesting to promote vegetative growth and flowering is desirable. All dead, diseased, crowded growth and suckers sprouting from the base and sides of the framework are pruned back annually. Pruned trees give larger fruits and early ripening. Guava is a axillary bearing crops, for inhancing the current season growth regular pruning is essential.

Flowering and fruiting: Under subtropical condition guava tree flower mainly in two seasons that is spring flowering commonly refffered as ambe bahar, for which fruit ripen in rain and rainy season flowering i.e. mrig bahar for which fruit ripen in winter. However, in tropical climate it is flowering three times and third flowering is in October i.e. hasth bahar for which fruit ripen in Feb-March.

Crop regulation in guava or bahar treatment in guava

As already stated that it is described to take only one crop in a year which give better and bulk production and high quality fruit, for achieve this the flowering and fruiting can be manipulated accordingly. There are various methods to regulate the bahar.

1. Hand de-blossoming.
2. The spray of growth regulator like NAA, at high concentration at full bloom period that is 400-800 ppm, 30 ppm have also found most effective.
3. Spray of high concentration of urea, (15-20%).
4. Spray of ethaphan 1000-1500 ppm at full bloom stage.

Disorder

Bronzing in guava : It is a complex nutrient disorder and this is due to P_2O_5, K_2O and Zn deficiency. It can be contol by soil application of diammonium phosphate in the month of December-January.

Insect and pest

Fruit fly (*Dacus dorsalis)*: Its attack mostly in rainy season crop. Female lays eggs under the skin of fruits from which maggot's imergese within 4-5 days, feed through their life of 6-20 days on pulp. The affected fruits rot and drop. It can be control by application of Malathion 0.05% or Labacid 0.05%.

Bark eating caterpillar (*Inderbela quadrinotata*): Bark eating caterpillar is an other important insect which is damage the bark of the trunck and branchese and their control is very cheap and easy. It can be control by kerosine oil or petrol.

Disease

Guava wilt (*Fusarium oxysporum* var. psidii): This is the most serious diseses in north India. Its found in alkaline soil the affected tree show slow browning of leaf which looks wilted and limb start drying up one by one and eventually whole tree dry. It can be control by-

1. Affected plant should be destroy
2. Plantation of guava should not be repeated in the same field.
3. Drenching of soil with 0.3% solution of thiram or 0.1% solution of benlet.

Anthracnose (*Collectotricum psidii*): This diseses is mainly affect the fruit during rainy season. In the beginning the rough spot appear on fruit surface which coalesce and become larger and on storage such fruits rot. This fungus also attack shoots flowers, buds and branchage begin to die-back. It can be control by spray of Bordeaux mixture 3:3:50 or 1-1.5% solution of Diphytolan.

Harvesting and yield : Guava, being a climacteric fruit, it ripens after harvesting; the fruits are harvested throughout the year (except during May and June) in one or the other region of the country. However, peak harvesting periods in north India are August for rainy season crop, November-December for winter season crop and March April for spring season crop. Guava fruit generally ripen 5 month after flowering and fruits are harvested when their colour change to yellowish green. Yield is about 150-200 kg per plant. In general in Prayagraj region 25 tons per hectare yield has been reported.

22

Pineapple

Botanical Name	: *Ananas comosus*
Family	: Bromeliaceae
Chromosome No.	: 2n = 50
Type of Fruit	: Sorosis
Edible Part	: Bract and perianth
Origin Place	: Brazil

Origin and history: Pineapple is considered as one of the most popular tropical fruit, and is known as "Golden Queen" all over the globe. It can be grown as mixed crop in most of the orchards and also as hedge plant and as a pure crop. It is drought tolerant and suitable for marginal lands. Pineapple contain an enzyme Bromeline and volatile compound three methyl propionate ester. It is widely cultivated throughout the world and country is like Malaya, Hawaii, Brazil, South Africa, Singapore, Mexico and India. In India it is cultivated in eastern part of the country like West Bengal, Assam, Manipur, Maharashtra, Andhra Pradesh, Goa, Madhya Pradesh and Tamil Nadu.

Composition and uses: Pineapple has been described as an excellent source of vitamin C and also good source of vitamin A and B, Sugar 13 per cent, Acidity-0.6-1.0 per cent Mineral matter-0.05 per cent, Fiber, 0.3 per cent, Ca, Fe, P.

- Fruits are relished as dessert in the form of slices either fresh or canned, preparation of juice, squash and jam and mixed jam. Candy is prepared from fruit core.
- Fruits are used in preparation of alcohol, vinegar, calcium citrate, citric acid, gum and pineapple flavours.
- The leaves yield silky fiber which is used for making a fine fabric known as Pina cloth in Philippines and Taiwan.
- The pineapple plants and the fruit residues after extraction of juice are used as cattle feed.

Soil and climate: Pineapple required sandy loam soil with pH range from 5.5-6.0. It is suitable to grow high rainfall and humid coastal regions and hilly area is not eastern region of the country. The optimum temperature is 22-32°C. Pineapple can be grown successfully from sea coast to an altitude of about 1500 m above sea level. The optimum rainfall ranges from 100-150 cm.

Propagation and type of planting material: Pineapple is a vegetative propagated crop and for which we can used several part of the plant like sucker, slip and crown. Sucker produce fruits 15-18 month after planting and crown produce fruit more than 24 month after planting. About 250-500 gm weight of suckers, 250-450g weight of slips and 40-45cm crowns are selected for planting.

Planting: Planting is done during kharif season, before planting suckers or slips should be sun-cured, dry leaves and scales at the base should be removed basal end of planting materials should be dipped in 0.4 per cent difolton and 0.05 per cent ekalux to avoid fungal & mealy bugs infestation. The suckers or slips are planted in 10-15cm deep holes, while planting, the growing part should not be buried. Planting may be done in single or double row systems. In single row system we can accommodate about 15,000-20,000 plants/ha. Where on the spacing 25 cm plant-plant, 35 cm row-row, 90 cm bed-bed, so the plant population is about 64000 plants/ha or some time spacing 25 cm x 35cm x 75 cm so the plant population is about 72727 plants/ha.

Manure and fertilizers: 20-25 kg FYM, 12-16 gm N, 2-4gm P_2O_5, 10-12 K_2O per plant per year. Fertilizers should be applied at 3 month of far planting in 4 split intervals. Foliar application of N (2-4%) is practiced.

Irrigation: Though pineapple is a rainfed crop, 4-6 irrigations may be necessary during dry season at 20-25 days interval.

Species: According to Smith (1979) established two genera, *Ananas* and *Pseudananas,* whereas *Pseudananas* at maturity bears a minute inconspicuous coma of bracts and plants produce elongated stolons and no slips, but *Ananas* bears conspicuous coma of foliaceous bracts and plant produce slips but not stolons and fruits remains seedless. The genus *Ananas* has 8 species viz.:

1. *Ananas comosus*: The cultivated species with large fruits and a short thick spike.
2. *A. monstrosus*: Crownless- There is no leaf formation on fruit.
3. *A. bracteatus*: Has well developed, bright red flower bracts. Fruit is edible and relatively large.
4. *A. Fruitzmuelleri*: It bears pale green, developed bracts.
5. *A. ananassoides*: With small fruits, erect leaves and a long fairly thin spike.
6. *A. nanus*: A dwarf species.
7. *A. parguazensis*: With curved leaves.

8. *A. lucidus*: It has smooth (Spineless) leaves from which goo quality fibre can be obtained.

But in case of genus *Pseudomonas* has only are species *viz. Pseudomonas* sagenarius.

Varieties: The varieties have been classified in to 5 groups.

1. **Spanish group:** Mauritius Red Spanish, Singapore Spanish, Green Selangor, Castilla, P.R.1- 67.
2. **Queen group:** Queen, Mac gregor, Natal, Ripley and Alexandria.
3. **Cayenne group:** Cayenne, Baron, Rothschild, Smooth Guatemalan, Kew, Giant Kew and Typhone.
4. **Abacaxi group:** Perola, Abakka, Sugar loaf, Papelon, Venezolana, Amarella.
5. **Maipure group:** Maipure, Bumuguesa, Rondon, Perolera, Monte Lirio.

Important variety: Giant Kew, Queen, Kew, Maurititius, Cayenne, Jaldhoop, Abakka, Red Spanish and Lakhath etc.

Flowering and fruiting: Pineapple flower 11-12 month after planting when plant attained 40 leafs some time flowering may be delay up to 15-18 month due to some environmental and nutritional condition. The flowering may in enhance by the application of ethaphon, aceteline, NAA, calcium carbide etc. flowering start in Feb-April, and fruit ripen in June-August.

Physiological disorders

1. **Sunscald:** The cells under the fruit skin of exposed surface to sunrays get damaged; care should be taken to control lodging and fruit should be covered with dry straw or its own leaves during April-May.
2. **Fasciation and multiple crowns:** Multiple crowns may occur due to genetical factor as well as due to soil and environmental reasons. It is also due to excess Nitrogen. Fruits gets flattened and fascinated. In fertile virgin soil of warm areas, more abnormal fruits occur as compared to less fertile soil.
3. **Black heart:** It is also known as endogenous brown spot or internal browning, formation of brown spots at the base of fruit-lets and further black discolouration of the centre core. Low temperature or exogenous application of GA_3 can induce this disorder. Fruits exposed to high temperature (40^0C) for 24 hours reduce black heart in cold stored pineapple.

Insect and Disease

Insect

Mealy bug (*Pseudococcus brevipes*): Control by-

1. Planted only the plastic material taken from healthy crop.
2. Treat the basal portion of sucker or slip by dipping in 0.04% solution of methyl parathion.
3. Apply thimate (soil born insecticide) @ of 10gm per plant at the 3 month interval. 10-12 kg thimate is sufficient for 1 ha crop.

Disease

Heart rot or tem rot (*Phytopthora parasitica*): Control by-

1. Provide better drainage facilities.
2. Select healthy planting material.
3. Treat the planting material in the solution of Bordeaux mixture (4: 4: 50).
4. Spray the crop with 0.1% solution of captan or Difolatan.

Fruit rot/soft rot/storage rot (*Ceratostomella paradoxa*): Control by-

1. Deep the sucker with the solution of Bordeaux mixture (2: 2: 50)
2. The cut end of the fruit stock should be deep in 10% solution of Benzoic acid with Alcohol.

Harvesting and yield: At maturity with a slight colour change at the base of fruit. The lowest eyelets have orange yellow colour and flattened. The bract becomes loose & brown. The fruits should be harvested along with 5-7 cm stalk on full maturity because it is non-climacteric fruit. The yield from a plant population of 350000-40000 per hectare is about 40-50 tones and that from a plant population of 43000-50000 per hectare normally varies between 50 and 60 tones. The average yield is about 600-1000 q/ha.

23

Aonla

Botanical Name	: *Phyllanthus emblica*
Family	: Euphorbiaceae
Chromosome No.	: 2n = 28
Type of Fruit	: Drupe
Edible Portion	: Mesocarp
Origin Place	: South Eastern Asia

Origin and history : Aonla or Indian gooseberry is also known as Indian amla. Aonla is the tropical South eastern Asia particularly central and southern India. It is commercially grown in Uttar Pradesh and Gujarat. In Uttar Pradesh aonla cultivation is more popular in Pratapgarh, Mirzapur and Barely. It is very rich from nutritional point of view being next to Barbados cherry in terms of vitamin C content (500-750 mg/100g pulp)which is not oxidized due to presence of tannins and even it is least destroyed during storage. Besides, it has an excellent medicinal value. It can be grown successfully in arid climate and in the soils of higher pH and poor fertility wastelands.

Tree is hardy and of medium height. However, seedlings tree can attain height up to 30m. it is evergreen, however, leaf shedding takes place during February-March. Tree trunk is smooth and whitish which resembles guava.

Soil and climate: Aonla can be grown in any type of soil but it performs well in sandy loam to clay loam. It has great tolerance to salinity and sodality and the pH range is 6-8 is very succesfully.

Its cultivation is more successful in subtropical regions. It can be grown successfully up to the elevation of 1800 m from sea level. It can be grown in subtropical and tolerant the freezing temperature as high as up to 46^{0}C. It is highly suitable for arid region. Some specific feature which are suitable for arid region plant :

1. Plant should have deep tap root system.
2. Plant should have waxi-coating on the arial part of the plant.
3. Plant should have capacity to bind the water (The maximum water binding capacity is in fig).

4. Plant should have characteristics shedding of leaf at the time of scanty sunlight.
5. The flower and fruit of the plant are shrinconized with the maximum ability of moisture in the atmosphere.

Important variety: Baranasi, Francis, Hathijhool, NA-4(Kanchan), NA-5(Krishna), NA-6(Chakaiya), NA-7(Amrit), NA-9 and NA-10.

Propagation: Aonla can be propagated by both sexual and asexual method. In asexual method budding particularly patch, modified, ring and forkert budding is very much succesfull. Seedling can be raised in seed bed or polethene bag. It will take about 4 month to attach budable size. The best time of budding is May-June but it can be done from May-June to September-October.

Soft wood grafting can also be used successfully for which seedlings are raised insitu and its head back in the month of May for forcing new vigorous side shoot. On these new shoots soft wood grafting method can be emplayed by way of wedge grafting.

Pruning and training: Aonla trees do not require special pruning and training. However, during the initial years pruning should be done to have balance training of the plant. In addition dried branches should be removed periodically.

Planting: The budded or grafted aonla plant is good for planting. The best time of planting is the begining of monsoon i.e. in the month of June-July. Distance of 8-10 m both ways is recommended for planting. The pits are made well in summer in the size of 1x1x1 m and pits kept open about one fourt night. Each pit have 3-4 baskets FYM mixed with soil. After first rain plant is planted in the center of the pit and staked properly.

Manure and fertilizers: One year old plant require 10 kg FYM and 150 gm N 100 gm P_2O_5 and 70gm K_2O per plant per year. For improving the fruit quality 0.6% borax should spray in the month of September-October at an interval of 10-15 days. Borax is also help full in controlling the fruit necrosis which is developed due to boron difficiency.

Irrigation: Aonla tree are hardy and stand very well against drought normally, 2-3 irrigation is beneficial at the time of flowering and fruit setting.

Flowering and fruiting: Flower bud differentiation in Banarasi aonla was observed during the first week of March and flowers started opening from the last week of March and blooming period is upto 3 week. There is no self incompatibility in aonla due to high percentage of staminate flowers. Flowers open in the morning whereas, dehiscence of pollen grain occurs in the evening between 4 and 6 pm. Only 30% pollens are viable in cultivar Banarasi and at room temperature these become non-viable within10 days. Stigmatic receptivity starts after three days of opening of female flowers and stigma remains receptive only for 18 hr. pollination occurs through wind as well as by honey bees.

The fruits remain in dormancy for about three and half months in cv. Banarasi dormancy discontinues from last week of July, 10th first week of august and fruits start developing gradually in the initial stage but rapidly during the month of September.

Disorder

Internal fruit necrosis (Blackening of Aonla): In this disorder necrosis symptoms start with the browning of inner most part of the mesocarpic tissue at the time of endocarp hardening in the second and third week of September. The browning of mesocarp extend toward the epicarp resulting in to brownish black are on the fruit surface in the second and third week of October. Its depending upon the seviourty of the disorder, mesocarp of affected fruit tern black and become corkay and gummy pocket develope.

This is due to deficiency of boron and it can be control by spray of 0.6% Borax thrice in the month of Sept-Oct at an interval of 10-15 days.

Insect, pest and disease

Gall catter piller (*Bentosa stytophora)*: Young caterpillar born into the apical portion of the shoot during rainy season and make tunnel, due to this damage reason swell out abrauptly in to a gall is provide space for movement of caterpillar. Due to this apical growth is checked. It can be control by pruning of the apical portion and spray of systemic insecticide like rogar 0.3%.

Bark eating caterpillar (*Inderbela quadrinotata*): Bark eating caterpillar is an other important insect which is damage the bark of the trunck and branchage and their control is very cheap and essy. It can be control by kerosine oil or petrol.

Diseases: Ring rust and fruit rot is a important diseases in aonla which can be control by 0.2% Dithane Z-78 at an interval of 7-15 days.

Harvesting and yield: Harvesting asexually propagated plant start fruiting as a commercial plant 6-8 year after planting. Generally aonla fruit are ready for harvesting in the month of Nov-Dec. normally a full grown aonla tree yielded 100-150 kg fruit /plant/year.

24

Litchi

Botanical Name	: *Litchi chinensis*
Family	: Sapindaceae
Chromosome No	: 2n = 30
Type of Fruit	: Nut
Edible Portion	: Flashy aril
Origin Place	: Southern China

Origin and history- Litchi is a popular subtropical evergreen fruit; the white translucent flavoured aril is liked very much and is used for table purpose in India. The fruits are available during May–June in northern India. China is the biggest producer of litchi followed by India. In India major litchi producing areas are Mujaffarpur, Bihar mountain area of Saharanpur, Dehradoon, Hoshiyarpur and Gurdaspur district of Punjab.

Soil and climate - It can be grown sandy to clay loam and it prefer slightly acidic soil having pH range from 6-6.5 it has been reported that in acidic soil litchi root develop Micorrhyza. Which symbiosis of root fungus and there by Nitrogen fixation take place in atmosphere.

It is a subtropical fruit required mild climate which is free from frost during winter and dry heat during summer. The hot desicating winds are the limiting factor in litchi cultivation.

Species: The family sapindaceae and sub family Nepheleae has about 125 genera and more than 1000 species. The genus Litchi has two species.

1. *Litchi philippinesis*: It is a wild type mostly used as root stock
2. *Litchi chinensis*: It is a commercial important species. The other members of the sub family are
   i. *Euphoria longana*- (Langan/Anshpal)- Small fruits, inferior quality.
   ii. *Nephelium lappaceium*- (Rambutan)- Hairy fruits.

Important variety: Mujjapharpur, Early Bedana, Saharanpur, Shahi, China, Kesba and Purbi are the important variety which is suitable for early crop.

Dehradoon, Saharanpur and Rose Scented are the important variety which is suitable for mid season. Calcatia, Seedless Late, Bombai, and Ellaichi are the important verity which suitable for late season.

Plant propagation: Air layering is the commercial method of propagation and best time for air layering is the beginning of monsoon. It can also be propagated by cutting budding and grafting.

Planting: Planting is done in rainy season or even just after the monsoon. The litchi trees are planted by following square system at 8x8m or 10x10m apart in 1x1x1 m pits.

Manure and fertilizers: One year old plant require 10 kg FYM., 100 gm N, 50 gm P_2O_5 and 100 gm K_2O and this dose is increasing up to 6 year and the 6 year old plant given 60 kg FYM, 600 gm N, 300 gm P_2O_5 600 gm K_2O.

Irrigation: First irrigation should be done just after transplanting. Irrigation is very important especially in summer (April-June) at an interval of 10-15 days and in winter crop should be irrigate at 15-20 days interval.

Training and pruning: Modified central leader system of training is followed and pruning is not done in litchi, except removing dried, diseased and criss crossed branches.

Flowering and fruiting: The asexually propagated litchi plant start flowering at the age of 3-5 year under good orchard management condition while a seedling tree take 10-15 year for first flowering. Flower start from last week of January to first week of February and fruit of ripen in May-June. In south India flowering start in the month of December-January and fruit ripen in April-May. Litchi is the cross pollinated and entemophyllus crop. The main pollinating agent is honey bee.

Physiological disorder

Fruit cracking: The fruit cracking in litchi is due to fluctuations in diurnal temperature, heavy irrigation/rain after prolonged dry spell, hailstorms during fruit development, boron deficiency causes injury to the fruit skin.

To avoid fruit cracking: The field to be irrigated during fruit growth and in early summer. Spraying with 2, 4, D (10 ppm), GA_3 (20 ppm) and Boron (0.4 per cent) is reduces the fruit cracking. Resistant cultivar like Dehradoon & Muzaffarpur should be planted in cracking prune area.

Insect and pest

Bark eating caterpillar (*Inderbela quadrinotata*): Bark eating caterpillar is an other important insect which is damage the bark of the trunck and branchage and their control is very cheap and easy. It can be control by kerosine oil or petrol.

Litchi mite *(Eriophyes litchi):* Due to seviour infestation of mite curling of leaves take place and it can be control by application by Dimethoate 0.05% on 3-4 spray.

Leaf roller: Due to seviour infestation on leaves 2-3 leaves are roll and insect are damage the leaves and it can be control by application of Malathion 0.2% and affected leaves are destroy.

Disease

Fruit rot (*Aspergillus resprictus*): This disease is most serious in storage and it can be control by application of difolton 0.2% solution.

Harvesting and yield: Litchi is non-climacteric fruit and harvesting is to be done at full mature stage. The tree starts bearing from 3-4 years after planting with proper care and management under suitable environmental conditions and the economic life of the crop maybe over 100 years. It takes about 55-60 days from flowering to harvest. The fruits are harvested during May and June. The maturity indices are flatness of tubercles and smoothness of epicarp and colour break down from green to pink. The whole bunch is harvested manually. About 80-150 Kg of fruits/tree in Indian conditions.

25

Banana

Botanical Name	: *Musa paradisiaca*
Family	: *Musaceae*
Chromosome No	: 2n = 22
Type of Fruit	: Berry
Edible Portion	: Mesocarp
Origin Place	: South East Asia

Origin and history : Banana is an important fruit of tropics. The fruit is recognized as the fourth most important global food commodity. India's share is 32 per cent of the total fruit production. This is the only tropical fruit which is exported in large quantities and is leading fruit in the international trade. It is one of the oldest and commonest of the Indian fruit that has been cultivated since ancient times. Eve was said to have used banana leaves to covers her modesty in the garden of paradise. Banana is thus called apple of paradise. It is also known as "Adam's fig or tree of wisdom". It is used as staple fruit in most of the African countries and is used as ripe (table) or raw fruit (cooking). This fruit is available throughout the year. All the plant parts are being used. Banana is monocotyledonous plant true stem is rhizome which is a modified form of stem fruit is seedless due to parthenocarpy and the development of fruit is due to auxin stimulation. India has more than 300 germplasms, out of 600 reported worldwide. Edible banana has arisen as a result of natural crosses between two wild progenitors viz,. *Musa paradisiaca* and *Musa balbisiana.* Banana is being grown in many of the African countries between 30^0 N and 50^0 S latitudes. Important countries are India (1st place).

Uses and composition: Banana by virtue of its, multiple uses is popularly known as "Kalpataru" (a plant with virtue). Usually banana is eaten when ripe as dessert / table fruit. Plantain or cooking bananas are the staple food of people in many countries of central and western Africa. Apart from fresh consumption, some types are also used for culinary purposes. The flower bud and also the central core of the pseudostem are used to prepare tasty dishes. Banana leaf is used as plates for serving food, leaf sheaths are used as wrapping material and dry leaves as fuel, while, tender pseudostems, leaves, underground rhizomes are used as cattle feed. Various processed products like banana chips, toffee, puree, powder, flour, vinegar, jam, jelly and wine can be prepared from the fruit. Banana fiber can be made in to attractive napkins, table mats and carry bags. The fruits have a lot of medicinal properties and are used for treating various health disorders specially acidity, ulcer, joint pains, high blood pressure and heart diseases.

Banana is rich source of energy (350 to 550 kilo joules/100g) and is a good source of minerals and vitamins. It contains 73 per cent moisture, 25-30 per cent carbohydrates, 1.4 per cent protein, 0.3 per cent fat, 0.5 per cent mineral matters (Ca, Fe, P, K) and vitamin- C and K.

Soil and climate : Banana can be grown in almost all types of soil provided adequate soil moisture is available. For successful cultivation banana soil should be deep well drained, friable with adequate organic matter. In alkaline soil reduce of banana wilt. Though banana requires large quantity of water, it cannot tolerate water stagnation. The optimum soil pH is 6.5-7.5.

Banana is a tropical crop requiring warm humid climate, it can be grown the temperature range for 10-40^{0}C and average temperature is about 23^{0}C.

Species and cultivars: The family musaceae has two genera viz., *Ensete* and *Musa* with about 50 species.

1. ***Ensete***- It is an old genus, which probably originated in Asia and spread to Africa. It has about 6-7 species, of which *Ensete* and *Ventricosa* has been reported to be grown as a food crop.
2. ***Musa*** –It is having 40-45 species, all the varieties under these species are under cultivation. The genus *Musa* is divided into following sections.

i. **Eumusa (2n=22):** This is the largest section with 13-15 species, all are cultivated forms; Pseudostem usually exceed 3m in height with pendent or semi pendent inflorescence, produces 10-25 nodes of flowers and covered with dull brown colour bracts, gives the edible cultivated parthenocarpic banana and are derived from 2 wild species i.e. *Musa accuminata* (A) and *Musa balbisiana* (B).

ii. **Callimusa: (2n=10)** -It is having 5-6 species. Plants with less than 3 m in height, suited as ornamental, parthenocarpy absent completely.

iii. **Australimusa (2n=20)-** These fruits are parthenocarpic and predominantly female sterile; The fruiting bunch is erect and contains a red sap, the skin is orange in colour when ripe e.g. *Musa textilis* (Manila hemp).

iv. **Rhodochlamys (2n=22)** -It is having 5-7 species, spreads from India to Indonesia, pseudostem less than 3m height with erect inflorescence, parthenocarpy absent e.g. *Musaornate* & *M. velutina* are sometimes grown as ornamental plants.

v. **Incertae sedis (x=7; 2n=14)** : It is the largest among the Musaceae family, grows to a height of over 10m. e.g. *M. ingens* & *M. beccarii* (x=9, 2n=18).

These edible bananas have 22, 33 or 44 chromosomes i.e., Diploids, Triploids and Tetraploids respectively. Triploid cultivars are generally numerous, diploids somewhat less and tetraploid forms are very rare. The basic haploid numbers is 11.

Simmonds and Shephered (1995) have distinguished the major morphological charecters of *M. acuminata* and *M. balbisiana*, which are as follows;

Characters	*Musa acuminata*	*Musa balbisiana*
Colour of pseudostem	Heavily marked with black or brown blotches	Blotches slightly or absent
Peduncle	Usually downy or hairy	Glabrous
Pedicel	Short	Long
Ovules	Two regular rows in each loculus	Four irregular rows in each loculus.
Bract curling	Bracts roll after opening	Bracts lift but do not roll
Bract shape	Lancedote or narrowly ovate tepering sharply.	Broadly ovate not tapering sharply
Bract apex	Acute	Obtuse.
Bract colour	Red, dull purpose or yellow outside, pink dull purple inside.	Brownish purple outside bright crimson inside.
Male flowers Colour	Creamy white	Variably flushed with pink.
Stigma colours	Orange or rich yellow	Cream, pale yellow.

The best known banana all over the world belong to the pure *acuminata* (AAA) group but the clones which are having both the parents is associated with the greater drought tolerance and resistance to diseases. e.g. AB, AAB, ABB, AA or AAA suited for rainy condition.

Genomic constitution of different cultivars of banana

In India bananas are distributed in southern, eastern, central and north eastern parts within 80^0 and 30^0N latitudes. Major genomic groups and cultivars are AA group:

1. AA- Anaikomban, Matti, Kadali, Tongat, Pisanglilin.
2. AB- Ney poovan (Elakki bale), Kunnan, Nathu Poovan. Thaen Kunnan, Adakka Kunnan, Ladyfinger, Safed Velchi.
3. AAB- Poovan, Rasthali, Pachanadan/Kaali/Galibale, Nendra Paditha, Rajapuri, Virupakshi/Sirumalai, Nendran/Rajeli, Chinali, Martmon, Hal Kela.

4. AAA- Dwarf Cavendish/Basrai, Giant Cavendish, Robusta, Gross Michel, Grand Naine, William, Nagabale, Chenkadali/Red Banana, Chakkarakeli, Amrit Sagar, Lal Velchi, Harichhal, Jayant Governor.
5. ABB- NallaBontha, Monthan/Kanchkela, Keribontha, Peyan, Karpuravalli, Sugandhi, Khasdia.
6. AAAA- Bodles Altafort, IC-2.
7. ABBB- Klue Taparod
8. AABB- Kalamagol
9. AAAB- Atan, Goldfinger (FHIA-1).

Evolution of edible banana- There are two wild species in banana i.e. *Musa accuminata* which is denoted by AA and *Musa balbisiana* which is denoted by BB. They are cross naturaly and making zygote AB in banana some varities are AB type. Due to result of poliploidy breeding make triploid and tetraploid variety. Tetraploid accuminata (AAAA) crossed with diploid bulbisiana and make AAB type (triploid) at the same time tetraploid bulbisiana cross with diploid accuminata and make ABB type (triploid).

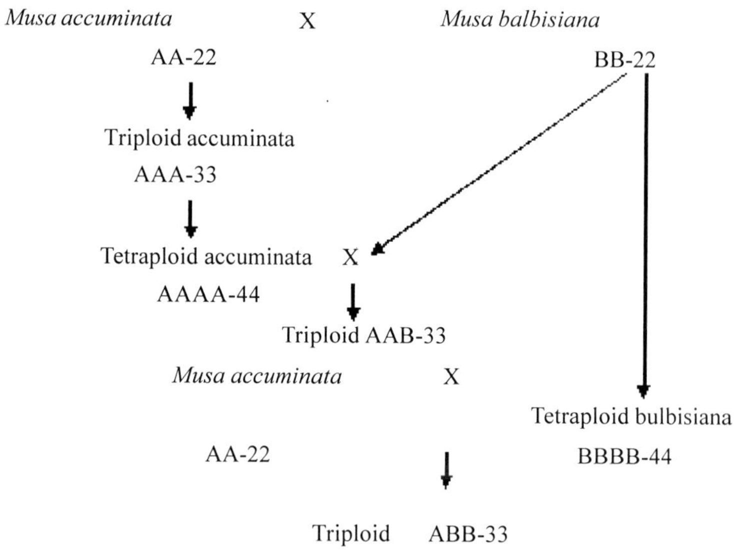

Some important hybrid which are also in cultivation

CO-1 : Kellar Laden X M*usa bulbisiana* X kadali.

H-1 : This is resistant to leaf spot fusarium wilt and Nematode.

F.H.A.I.-1 : (AAAB) (Gold Finger): Resistant to panama wilt and sigatoka leaf spot, gher weight 18-20 kg. It is belong to pome group.

Bodles Altafort : Synthetic hybrid variety, a result of cross between Gross Michel (AAA) X Pisanglin (AA).

Padarshe (Hanuman): It is an average bunch weight 55-60kg.

Klueteparod (AABB): It is natural tetraploid.

Propagation: Banana is commercially propagated by rhizome (sucker). In banana there are two type of sucker.

1. Water sucker.
2. Sward sucker.
3. **Water sucker**: Water sucker is one which is characterized by broader leaves which do not produce a healthy banana clump, with slender pseudostem. Rhizome/corm is not well developed. It takes more time (more than 18 months) for yielding and yields also less. These types normally develop from shallow buds away from pseudostem near the soil surface.
4. **Sword sucker**: Sword sucker is one with well developed rhizome, well developed pseudostem with sword like leaves. It takes 12-13 months to yield and gives bigger bunches. Sword suckers are closely associated with the mother plant and therefore develop strong thick rhizomes of their own.

The sward sucker have narrow and slender leaf where as the water sucker have broader leaf the flowering and fruiting always take place in sward sucker and there is no flowering and fruiting in water sucker.

Important criteria have to select the suckers for planting

1. The orchard/mother block should be disease free.
2. The weight of the suckers should be 1.0-1.5 kg.
3. The mother plant should be heavy yielder.
4. Always select sword suckers for planting.
5. Select the suckers free from rhizome weevils and age of suckers 3-4 months.

The whole or split rhizomes can also be used when suckers are not available. Bits of rhizomes may also be used as a planting material.

Selection of planting material

Fallowing point are keep in mind at the time of selection planting material.

1. Sucker should be 18-20cm in height and 1-2kg in weight.
2. Sucker should be narrow leaf like sward.
3. Base of the sucker should be thick and tapering toward the top.
4. Sucker should be always free from disease.

Planting: Banana can be planted throughout the year except in severe winter and during heavy rains. In general, September-October and when irrigation facility is proper, then planting should be done in February – March. In hilly area it should be done in April.

1. **Pit method** : Pit method and furrow methods are commonly followed. The pit size of $60cm^3$ should be opened at 1.8 x 1.8m or 2 x 2m (Tall varieties) adopting square system. These pits are filled with top soil with 20-30 kg. FYM should be applied at least 15-30 days prior to planting. During planting each pit will be supplied with 250gm neem cake and 50 gm of trichoderma to prevent nematode & rhizome rot problems. Planting of suckers at the centre of pit and irrigate immediately after planting. Spacing should be adopted variety wise.
2. **Furrow method:** This is the most common method of planting. Furrows of 15-20cm deep are opened at a regular distance and rhizomes are planted in the furrows. Paired row planting in tissue culture plants.
3. **Tissue culture plants:** Banana is also grown commercially by using tissue cultured plants these plants required much care throughout the growth period compare to suckers and yields about 10-20 per cent more than suckers.

In recent years the concept of HDP is being practiced, suckers are planted at closer spacing or planting two suckers per pit by accommodating more number of plants at specified spacing to get higher yield and reduced cost of production. The cultivar Robusta and Dwarf Cavendish spaced at 1.5x1.5m accommodates 4444 plants/ha is recommended by IIHR was recorded highest yield.

Manure and fertilizers: Banana is a heavy feeder crop, due to shallow root system of the crop, it responds well to applied nutrients. The high fertilizers requirement of banana is mainly due to their rapid and vigorous growth and high fruit yield. Generally one year old plant require 100-200 gm N, 100-150 gm P_2O_5 and 200-300gm K_2O per year.

Irrigation : Banana is a moisture loving plant it requires more irrigation. Irrigation should be done at 10 days interval from December-June. Generally 40-45 irrigation are required from planting to harvesting.

Inter culture operation

Desuckering: Desuckering is done by cutting the pseudostem of sucker at the ground level followed by application of kerosene/2-4, D @ 0.5 per cent also the growth of suckers can be inhibited by damaging the cut end to prevent further growth. As banana produces number of suckers, if allowed, they compete for moisture nutrition with mother plant; one or two healthy sword suckers may be allowed to grow for ratoon crop.

Propping : Providing support to the plant when it is at bunching stage. If dwarf variety and closed spacing no propping is required but for all tall varieties, it is required by using bamboo poles or any other supports.

Earthing up: To prevent uprooting of plant by wind soil is mounded around the pseudostem during rainy season.

Denaveling: Removal of male buds after the last set of fruit. It increases the bunch weight/fruit weight and quality of fruits also.

Thrashing: A process of removal of old, dry, diseased and senescent leaves this could reduce the disease and facilitate better light, temperature and air. However, if leaves are pruned before bunch initiation, flowering is delayed and cycle time increased. A minimum of 12 leaves are required to be retained for maximum yields

Wrapping : It is the process by covering of bunches with the help of polythene or gunny cloth, to protect the fruit from intense sun light, hot wind etc.

Bunch covering: Bagging of bunch with perforated polythene cover or dried leaves to protect against cold sun scorching, attack of thrips and other scrapping insects, during bunch maturity stage, the bags may be coated with pesticides.

Mulching : The covering of the soil with organic or inorganic material for conserving the soil moisture.

Removal of unwanted plant : In banana plant developed by water sucker should be removing.

Mattocking: It is the process of cutting the pseudostem after harvesting of bunches. After harvesting the pseudostem should be cut leaving a stump of about 0.6m hight, the left over stump with its stored food material continues to nourish the daughter sucker (follower) till it withers and dries up.

Flowering and fruiting : The inflorescence of banana plant emerges 9-12 month after planting and fruits are ready to harvested 3-4 month.

Insect and pest

Stem borer (*Odioporus longicollis*) : Stem borer is the most harmfull insect in banana cultivation this insect is bore the stem and after seviour infestation leaves become dry and ultimately plants are died. This insect can be control by application of Indo-sulfon 0.04% or Carbaride 0.1% and the infested plant should be removed from the field.

Root stock weevil (*Cosmopolitus soridisus*) : This is also a serious pest in banana and fruit become undersized. This is control by spray of Phaphamedan 0.3% or Fenetrothian 0.05%.

Banana aphid (*Pentalonia nigronervosa*) : Banana aphid is the serious pest in bananan cultivation this is found near the root zone of banana this insect suc the sap of the younger leaves and this is also transmiting agent of bunchy to virus. It can be control by spray of Rogar 0.3% or Phaphomedon, Monocrotophos 0.03%.

Disease

Banana wilt/Panama disease (*Fusarium wilt*)- This disease was first reported in panama in 19th century. The casual fungus is soil born get entry in plant body the root. Panama disease is most serious is poorly drained soil which planting of banana year by year. It is serious in poorly drained soil. Resistant varieties are Robusta & Dwarf Cavendish.

Control

1. Up rooted the different plant and destroy them.
2. Avoid banana planting in infected soil.
3. Use disease free sucker for planting.
4. Apply Bavistin @ 1gm/litter of water.

Banana bunchy top virus (BBTV) : This disease of first reported is fizzy in 1891 and transmitted by aphid vector, *Pentalonia nigronervosa.* The dwarf banana cultivars are very susceptible. The leaves are bunched together like a rosette at the top the margins are wavy and slightly rolled upward. Dark green streaks of the lamina or midrib.

Control

1. Removal of effected plant.
2. Planting material should be virus free and always taken from a reliable force.
3. Control the aphid by spray of 0.3% Rogar.

Leaf spot/Sigatoka : It is a fungal disease, initially, presence of light yellowish spots on the leaves under sever condition formation of brown spots and later dies, turning light grey surrounded by a brown ring. The Gross Michel and Cavendish group are all (AAA) highly susceptible to sigatoka while, all ABB clones are resistant.

Harvesting and yield : Banana is a climacteric fruit it should be harvested when they are green and fully mature. The sign of maturity in banan is fruits are harvested when tops leaves are start drying, the colour of the fruit change from deep green to light green, the size of the fruit and fruits loss their angularity.

The well managed crops of dwarf variety yielded 300-400 quintal per hectare whereas, the tall variety yielded 150-200 quintal per hectare. The yield of dwarf variety is higher due to more plant population per unit area.

26

Papaya

Botanical Name	: *Carica papaya*
Family	: Caricaceae
Chromosome No	: 2n =18
Type of Fruit	: Berry
Edible Portion	: Mesocarp
Origin Place	: Tropical America

Origin and history: Papaya has long been known as wonder fruit of the tropics and is one of the five major fruit crops of tropical world, papaya fruits throughout the year and a remunerative fruit crop which is grown in almost every part of the country. It gives maximum yield per unit and income next to banana. It is generally grown as fruit plant in kitchen garden, home garden or nutrition garden and later as commercial fruit. It is also planted as a filler plant in a fruit orchard. India is the second largest producer of papaya in the world after Brazil. It is a fruit native to tropical America and commercially grown in many parts of the world viz., Australia, Hawaii, Taiwan, Puerto Rico, Peru, Florida, Texas, California, South Africa, Pakistan, Bangladesh and India. In India, it was introduced by Portuguese in 1611. It is grown in Karnataka, Bihar, Kerala, Tamil Nadu, Assam, Gujarat, Maharastra, Andhra Pradesh, Madhya Pradesh etc.

Composition and uses: Papaya is a very delicious, wholesome, refreshing fruit with a unique food value. The ripe fruits are used for desert purpose, preparation of jam, jelly, soft drinks, ice cream, flavoring, crystallized fruits, canned in syrup, fruits salad. The unripe fruits are consumed as vegetables also as a contraceptive. The seeds also have medicinal properties.

The milky latex which is obtained from immature fruits is known as papain, which is a proteolytic enzyme used in several ways.

- As a digestive aid and for the treatment of ulcer & diphtheria.
- In the industry of pre-shrinking of wool and degumming of natural silk and rayon.

- Tenderization of meat and it is used for tanning of leathers.
- Used in brewing industries as a clarifying agent for beer etc.
- Used in preparation of drugs for treating intestinal cancer, tape worms, round worms and kidney disorders. The carpaine alkaloid present in green parts and in seeds used as heart stimulant, as diuretic and for treatment of amoebic dysentery. Preparation of chewing gum, cosmetics, papers and adhesive materials.
- It is also be used for extraction of oil from liver of Tuno fish.

Papaya is very nutritious, it contains high quantities of vitamin A and ranks second to mango in the content of vitamin-A (carotene), the vitamin A is associated with carotene but the yellow pigment of papaya is caricaxanthene. It also contains vitamin C, riboflavin, niacin, good source of Ca, P, Fe, fat and carbohydrates. Apart from papain, raw fruits can also be used for extraction of pectin. Pectin is generally extracted using alcohol precipitation method. It is used in food industry as flavouring extract and emulsifying agent.

Tutti frutti: Papaya bits (immature) are first soaked in brine and boiled in sugar syrup and immersed in sugar syrup for 48-60 hours with permitted colours. The other products like, papaya jam, canned slice, papaya beverages, fruit bars/ toffee bars, soft drinks etc, could be made out of the fruits.

Climate and soil- Papaya is a tropical fruit however, it also grows well in mild subtropical climate. Temperature is one of the most important climate factore. Which is determined the success of papayas cultivation. It is very much sensitive to frost strong wind and water. It is susceptible to frost. It requires worm humid climate and can be cultivated up to an elevation of 1000-1200 m above MSL. It thrives well at temperature between 38-44^{0}C but optimum range is 22-36^{0}C a day temperature of 35^{0}C and 25^{0}C night temperature are most suitable.

The well drained sandy loam soil is ideal for its cultivation. Soil with high pH and low pH should be avoided. Papaya grows well under well drained soil with a pH range of 6-7. It prefers fertile soil, deep clayey and calcarious and rocky soils are not suitable.

Species: The genus *Carica* contains 48 species of which, only 3-4 species are important.

1. *Cariaca papaya* – It is a cultivated commercial species and polygamous in nature.
2. *C. monoica* – Monoecious and is suited for mild climate, found in Amazon basin.
3. *C. candamarcensis* – This species is known as mountain papaya is suited to cold climate.

4. *C. gracilis* - Ornamental
5. *C. pubescens* - Mountain papaya
6. *C. dodecaphylla* - Jaracatia
7. *C. goudotiana* -Papayuelo
8. *C. stipulata* - Chamburo
9. *C. quercifolia*- Oak leaved papaya

Important variety : Papaya is a polygamous crop according Storey. In Papaya 32 type of sex form has been reported. Some of the important sex form and there verities are as fallows.

PUSA, Bihar-released some improved varieties.

a. **Pusa Delicious** - It is a gynodioecious with medium sized fruits (1.5kg) and high yielder, male has been eliminated genetically by sibmating with hermaphrodite.

b. **Pusa Majesty**: A gynodioecious, medium round fruits with good keeping quality. Plants resistant to virus diseases. The fruit with stand long distance transport.

c. **Pusa Giant**- Dioecious variety with big sized fruits (2-3.5 kg), plants are very vigorous and resistant to wind damage. The fruits are suitable for canning industries.

d. **Pusa Dwarf**: The plants are dwarf and bears fruits at 38cm from the ground, they are dioecious and medium yielder, medium sized fruits, suitable for HDP and home gardens.

e. **Pusa Nanha**: A dioecious dwarf mutant and well suited for kitchen garden, nutrition garden and pot cultivation. It yields about 60-65 tonnes of fruit/ha. It is developed by gama radiation, (Suitable for high density orcharding) this variety is developed by Prof. Mansha Ram.

a. **Dioecious varity:** CO-1, CO-2, CO-4, CO-5, CO-6, Pant Papaya-1,2,3 and Honey Dew (Madhu Bindu) etc.

b. **Gynodioecious varity:** CO-3, Coorge Honey Dew, Pink Flesh Sweet, Sunrise Solo, Taiwan, Washington, Surya- hybrid (IIHR-39) (Sunrise Solo x Pink Flesh sweet) etc.

Propagation : Papaya is commercially propagated by seed although cutting and budding also found succesfull in papaya. Papaya is a cross pollinated crop. Collection of papaya seed is the major factor in papaya production. Papaya seeds are extracted from mature fruit at 120-130 days. Freshly extracted seed with the sercotesta intact has low germination due to presence of inhibitor therefor seeds are properly cleaned to minimize the effect of inhibitor. Therefore cleaning of papaya in sawdust and drying in shed give better germination. The optimum temperature for seed germination in papaya is 35^0C. Germination is enhance by seed treatment with GA_3 100-120 ppm or thiourea 100 ppm. The seed rate is about 250-300gm/ha and seed bed size about 3 metre length 1 metre width and 10cm about the ground level. Nursery bed should be treated with the help of formaldehyde 2.5% and cover with polythene if disease appears in the nursery Bordeaux mixture 1% or Copper oxy-chloride 0.2% should spray.

Sowing: Seeds to be sown in poly bags (22x15cm with 150 gauges) 6 weeks prior to transplanting. The media should be disinfected by using 2 per cent formalin solution, 4 seeds per polybag should be sown, if it is dioecious variety or 2 seeds, if it is hermaphrodite variety. Germination will be observed in 2-3 weeks after sowing. About 6-7 weeks old seedlings will be ready for transplanting.

Planting : The papaya plant can be planted three season in a year the spring planting can be done in the month of February-March where climatic condition is mild throughout year and the monsoon crop can be planted in June-July. This planting is more common in frost proone area. The autumn planting can be done in the month of Oct-Nov. where rain fall is very high and virus problem is acute. The pit size is 60 x 60 x 60 cm should be dug at planting distance 1.8 x1.8 m or 1.3 x1.3 m or 1.4 x1.4 m or 1.6 x1.6 m or 1.2 x1.2 m (Pusa Nanha) and planting should always being done in evening period.

Manure and fertilizers: At the time of filling of pit 25 kg FYM, 1-1.5 kg neem cake, 1.5-2.0 kg bone meal should be added. For one year old plant 10kg FYM, 100gm N, 50gm P_2O_5, 100gm K_2O should be given. The application of nutrient at bi-monthly interval is recommended.

Irrigation : The soil moisture is important factor which affect the survival of the plant, growth sex change fruit yield and quality of papaya the total water requirement of papapya under tropical region is estimated to be 1800-1900mm. excess application of moisture causes reduce the growth and yield. In general in papaya irrigation should be done at 15-20 days interval in winter and 8-10 days interval in summer. The low level of moisture at the time of flowering and fruit set causase sterility and more number of male flower.

Removal of the unwanted plant : In case of diocious variety the male plant should be removed and 10% male plant should be left for proper pollination. In case of gynodioecious variety there is no need of any pollinizer.

Flowering and fruiting: Papaya is a polygamous crop. The sex expression of papaya under the cultivation determine the pollination fruit set and ultimately the fruit yield. Generally there are three sex form in papaya i.e. pistillate, hermaphrodite and staminate. Pistillate flowers is unisexual and stable sex form which is not modified by any environmental factor in between male, female and hermaphrodite. Several variation of basic type are recognised which affect the different degree of maleness and femaleness. In papaya their are five main type of flower are recognised which are as follows.

1. Pistillate
2. Pentandria (pistillate having five stamen attach to ovule)
3. Intermidiat (unstable)
4. Ellongated (hermaphrodite)
5. Staminate (flowers on the long pidauncle)

In papaya the 32 type of sex form has been reported which is confirm by Pusa Institute, which is manifested depending upon cultivar and season. Papaya starts flowering in 5-6 months after planting, when the sex of the plant is identified, surplus male plants (plants with long 1-1.5 meters flower stalk hang out with small tubular flowers contains stamens only), may be removed keeping one male plant for every 10 female plants, as pollinizers and removing of male plants by maintaining male plant for every 10 plants after flowering.

Identification of sex in papaya: Attempt have been made to identified the sex of papaya plant before flowering by using seed colour, vigour of seedling chemical constituent of the plant, protein, amino acid and phenolics etc. In Honey Dew variety the dark seed colour was associated with the high proportion of male or hermaphrodite flower. The dark brown seed of medium size is pistillate flower whereas, the dark brown seed of large size produce predominantally hermaphrodite flower. The reproductive tissue of male plant shows greater activity of acid and alkaline phosphate than those of female plant.

Effect of growth regulator on sex expression: The several plant growth regulator have been suggested the change of sex in papaya. Spray of 2, 3 dichloroisobutarate on hermaphrodite form of dwarf papaya increase the number of female flower. Spray of GA_3 at 50 ppm increase the female flower in CO-1 variety of papaya. The application of SADH (2 succinic acid dimethyle hydrazide) at 250 ppm and phosphone-D at 250 ppm increase the femaleness in CO-1 variety. Similarly increase in femaleness was reported with TIBA at 25 ppm.

Effect of environment on sex expression: The long day treatment with a photoperiodic cycle of 16 hr. light and 8 hr. darkness favours the maleness in CO-1 variety.

Papain production technology: The green papaya fruit is well known for abundance of milky latex containing papain which have both milk clotting and protein digestive properly. Papain is a typical protein digestive enzyme it classed as protieas and protienase enzyme.

Usese of papain: Papain is used as an essential ingradiant in pharmaceutical, textiles, tanine industries and papain also use in tenderizing the meat, clearness of bear, manufacturing of cosmatics, dental pest and degumming of silk and rayon in tanin and lather industries. Papain is also used for treatment in stomach and intestinal cancer and spinal card.

Papain harvesting technology: Following points are keep in mind at the time of papain harvesting.

1. **Selection of the fruit:** The fuly mature oblong to cylindrical fruit of pistillate plant yield more papain. The optimum age of the fruit found to be 70-110 days after fruit set.
2. **Season and time of papain harvesting:** Time after 11 am and sesson are the month with temperature below 10 ^{0}C should be avoided. Therefor the fruit shoud be tapped for the papain in the month of October – January.
3. **Treatment for increasing the papain yield:** Growth regulators like 2,4 D 1 ppm, 2,4,5-T 25 ppm at blooming stage and NAA 75 ppm at preblooming stage have been reported to increase the papain yield and 1-2 % urea and ethephon have also been known to give more papain yield.
4. **Harvesting methodology:** Papain from selected fruit is harvested by injury on pericarp or fruit surface which is known as tapping. It should only done by stanless stainless blade which is mounted on a stick. Latex oosing from the fruit is collected in arecanut spath, glass vassels, aluminum or stainless steel tray which is coverd by the moshline cloth. The tray should be mountain on which have clamps with tree trunck at determinate hight. The papain harvestors advise to use rubber hand glubs to protect the hand from irritant action of latex. Fruits are retapped at an interval of 12-15 days until fruits are starts ripening.
5. **Papain processing:** The collected latex is passed through 10 mesh sieve and then added either 0.5% potasium metabisulphide and 0.2% thymer or sodium bisulphate single or incombination within 4 hrs. of its collection and prior to drying increase the storage life.
6. **Papain dehydration:** Drying of latex can be done at temperature range of 37-60^{0}C but it prefer optimum temperature for latex drying is 50-55^{0}C.
7. **Packaging and storage:** High grade papain should be pure white, granular, free from bad odder and of high proteolytic activity. Papain can be packed in air tight teen container with wax coatting but it should preferably repack in glass container or poly bag.
8. **Papain yield:** A well managed crop yielded 5-8g papain per fruit. Its varies from variation of climatic condition and temperature. Normaly 30-46g papain yielded per plant and 27.24-79.49kg per acre per annum.

Insect and pest

Aphid (*Myzus persicae*) : This is the sucking type pest which sap the leaves and transmit the viruses from one plant to other plant. It can be control by application of 0.5% monocrotophas or rogar.

Red spider mite (*Tetranycous cinnavarinus*) : This is a sucking type pest and suck the sap from younger fruits in seviour infestation and it can be control by application of 0.04% dicofal.

Disease

Collar rot (*Pythium aphenitematum*) : It is soil born fungus and in this disease water soaked patch are found on the stem which gradually in large and girdle base of the bark finally the affected portion are black and rot. It can be control by the application of copper oxycloride 2g/liter of water which check the further spread of desiese.

Root rot (*Phytopthora palmivora, Rhizoctonia solani* and *Fusarium species*): This disease can be control by seed treatment with fungicide like capton or cereson, sterilization of nursery bed by burning dry leaves, spray of formaldehyde two week before sowing and drenching the soil with copper fungicide.

Anthracnose (*colletotrichum gloesporiodes*) : The initial symptoms are small round water soaked area on fruits which later develop into sunken or concentric lesions. The disease also affects the petioles of lower leaves leading to their shedding. This can be control by broad spectrum fungicide like bavistine @ 2 g/ liter water.

Powdery Mildew (*Oidium caricae)* : The fungus is found mostly growing on the upper surface of the leaves with drawing nutrients from the cells of the leaf surface. Under severe attack the top portions of the seedling may die. This can be control by broad spectrum fungicide like bavistine @ 2 g/liter water.

Papaya mosaic : The affected plant are stunted in growth, yellow mottling, distortion of leaves, bending down the petiol followed by death of the plant. Diseased plant should be erradicated and spread by aphid and control by the application of any systemic insecticide @ 2-5 ml/litter of water.

Papaya leaf curls : This is viral disease and it is characterise by the seviour curling and distortion of leaves accompanied by reduction in leaf size. This virus is transmited by *Bemissia tobaci*. It can be control by roguing the affected plant at the early stage of disease development. The trasmiting agent can be control by application of any systemic insecticide @ 2-5 ml/litter of water.

Harvesting and yield : A mature fruit ripened when their colour change from geen to yellowish green, when latex of the fruits become almost watery. The fruit is considerd to be reddy for harvest in 12-14 month after planting. On an average the each plant of the improve variety bear 30-45 fruit weighing 40-75 kg in one fruiting season on an average 60-75 ton fruit/ha, in case of high density orcharding (Pusa Nanha) fruit yield can be obtained upto 60-65 tone per hectare.

27

Citrus

Botanical Name	: *Citrus* spp.
Family	: Rutaceae
Chromosome No.	: 2n=18
Type of Fruit	: Hesperidium
Edible Portion	: Juicy placenta
Origin Place	: Asia

Origin and history: Citrus is the leading fruit crop in the world. In India, citrus ranks second in area sharing 12.8% of the total area under fruit crops with 9.6% share in production. Under citrus group, mandarins are the most important, occupying 50% of the total area under citrus followed by sweet orange and limes. Citrus fruits are grown mainly in Maharashtra, Andhra Pradesh, Punjab, Karnataka and North-Eastern region.

1. Nagpur mandarin was introduced in India in 1894 by Shriji Raja Bhosle.
2. Kinnow can be grown in high density planting by using the cultivar Troyer Citrange' as a root stock by spacing the plants at 1.8 x 1.8 m^2.
3. Classification of citrus was given by Tanaka and Swingle.
4. Grape fruit is also known as Forbidden fruit, Breakfast food.

Citrus native to South East Asia, tropical and sub-tropical regions of Indio-China. Major citrus growing countries are USA, Spain, India, Italy, Japan, Argentina, Mexico, Brazil, Morocco, Algeria, Greece, South Africa, Australia, Israel, Egypt, Jamaica, etc.

Soil and climate: Citrus can be grown in a wide range of soil but thrive well in deep, loose, well aerated sandy loam soil. The ideal pH of citrus cultivation is 5.5-7.5.

Citrus belonging to the subtropical climate but thrive well in frost free area. The climatic factors like light and temperature are much important in citrus cultivation. For good quality fruits citrus require dry semi arid climates. It can tolerate occasional light frosts. Optimum temperature range is 16-20^{0}C within a range of 17-40^{0}C and annual rainfall of 500-775mm is optimum.

Species and cultivars: The classification of citrus was done by two famous authorities on the subject. W.T. Swingle (USA) and T. Tanaka (Japan) are at two extremes.

- Swingle (1948) recognized only 16 species under the genus *Citrus* where as
- Tanaka (1954) described as many as 144 species. He failed to cover many forms of horticultural importance and many species of Japanese, Chinese and Indian origins have been denied.
- Swingle divided the genus *Citrus* into subgenera viz; *Eucitrus* having 10 species and *Papeda* having 6 species.
- Tanaka's (1954) treatment although considered more comprehensive and detailed, contained excessive number of species, some of them being of doubtful validity. The mandarin group alone, he described 35 species, resulting into much confusion and obviously to lesser practical utility.
- Tanaka divided the genus *Citrus* into two subgenera viz; *Archicitrus* having 98 species and *Metacitrus* with 46 species.
- Contrary to Swingle's opinion, citrus forms of hybrids and certain cultivars by Tanaka were questionable and might be avoided. It is accepted that the characters employed for identifying a valid species of citrus should be free from the environmental influence.

Major species and varieties

1. Mandarin group

Citrus reticulata: It is Chinese origin place fruit crop. Polyembryonic cultivars are Nagpur Santara, Coorg Mandarin, Khasi Mandarin, Kinnow Mandarin, Sikkim Mandarin of India and Ponkan of China.

C. unshu: Japanese origin, seedless, cultivars are Satsuma Mandarins of Japan and Owari, Kara, Silver Hill, Ikeda Mikada.

C. deliciosa: Mediterranean origin, Polyembryonic cultivars, Willow leaf Mandarin, Kinnow, King of USA and Blinda of Algeria.

C. nobilis: Indo-China origin, natural Tangor. Polyembryonic cultivars like Kunembo of Japan, King Orange of USA.

2. Orange group

C. sinensis: Sweet orange polyembryonic cultivars are Mosambi, Maltablood Red, Sathgudi, Valencia, Pineapple, Washington Navel Orange, Shamouti of

Israel, Succari of Egypt, Dobla Fina of Spain, Mudkhed (bud mutant of Nagpur mandarin)

C.aurantium : Sour orange, Bitter orange, used for pickling purpose

3. Grape fruit and pummelo group

C. grandis (Pumelo): Monoembryonic, Malaysia and Polynesia origin. Leaves pubescent in lower surface, fruits in clusters. Cultivars are kaopan of Thailand, Buntan of Formosa, White fleshed fruit and Red fleshed fruit.

C. ***paradisica*** *(***Grape fruit):** South China origin, polyembryonic. Leaves non-pubescent, fruits solitary. Cultivars are Poser, Ruby, Marsh, Duncan Seedless, Thompson, Red blush, Triumph, Sharanpur Special.

4. Acid lime group

C. limon **(Lemon)**: Weakly polyembryonic, cotyledons white. Cultivars are Eureka, Lisbon of USA, Feminello and Monactiello of Italy, Bernia of Spain, Pant lemon-1, Baramasi, Assam lemon and Galgal of India. Lemon oil is very important.

C. jambheri **(Rough lemon)**: Polyembryonic Indian origin, cotyledons light green, popular rootstock, fairly tolerant to virus diseases.

C. aurantifolia **(Acid lime or sour lime)**: Polyembryonic, cotyledons whitish-popular cultivar are Kagzi lime-susceptible to tristeza and canker, Pramalini, Vikram, PKM-1 and Seedless of India.

C. limotoides **(Sweet lime)**: Major cultivars are Mitha Nimbu, Sharbati.

C. medica **(Citron)** : Indian origin, Monoembryonic, Persisting style.

C. karna **(Kharna Khatta)**: Popular rootstock, cotyledons white.

C. limonica **(Rangpur lime)**: Hardy nature, popular rootstock, tolerant to tristeza and also salt.

Other related wild species are

C. indica- Indian wild orange with inedible fruits

C. latipes- Khasi papeda- cold tolerant.

C. macroptera – Melanasian papeda- has medicinal value

C. ichangensis- Ichang papeda- cold hardy, fruits inedible.

C. assamensis- Admajor (Gajanimbe)

Related genera: *Ponicirus*- (Trifoliate orange) Fruits inedible, Polyembryonic rootstock.

Fortunella (Kumquat): The species are margarita, japonica, errasiflora, hindsii, Polyembryonic plants ornamental with small oval fruits.

Intergeneric hybrids

1. **Citrange-** Trifoliate orange x *C. sinensis*, cultivars are Troyer, Carriyo, Morton, Stonia, Rusk, Coleman.
2. **Citrange quat-** Trifoliate orange (*Ponicirus x Citrus*), citrange x fortunella-trigeneric hybrid (kumquat).
3. **Citrangedin-** (*Ponicirus trifoliate* x *C. sinensis*) x *C. mitis* (calamondin) – bigeric hybrid.
4. **Citrangor-** Citrange x *C. sinensis.*
5. **Cicitrange-** Citrange x *Ponicirus trifoliate.*
6. **Citrandarin-** *P. trifoliata* x *C. reticulata* (mandarin).
7. **Citrumelo-** *P.trifoliata* x *C. paradisi* (grape fruit).
8. **Citermon-** *P. sp* x *C. aurantium.*
9. **Citrumquat-** *P. sp* x *C. japonicum* x *F. margarita* (kumquat)

Intrageneric hybrids

1. **Tangor** : *C. reticulata* x *C. sinensis*, cultivars are Temple, Clamentine, Montreal, Umatilla, monoembryonic.
2. **Tangelo:** *C. reticulata* x *C. paradise*, cultivars are Orlando, Sampson, Seminole.
3. **Lemonima:** *C. limon* x *C. aurantifolia.*
4. **Lemmonnage**: *C. limon* x *C. reticulata*

1. Sweet orange

Second largest citrus fruit in cultivation and commercially grown in Andhra Pradesh (Ananthpur, Cudappah, Nalagonda, Mahaboobnagar and Chittor district), Maharasthra (Marathwada, Ahmednagar, Pune and Nasik), Karnataka, Punjab, Haryana, Rajasthan.

Varieties: Blood Red in Haryana, Punjab and Rajasthan-Jaffa, Hamlin, Pineapple- exotic, Mosambi in Maharasthra and Sathgudi in Andhra Pradesh.

Propogation: Citrus can be commercially propagated by seed. In asexual propagation T-budding is more common in sweet orange, mandarin, pummelo etc. and the best time of budding is January-March or September-October. In lemon and sweet lime air layering is commercially practiced. Jatti Khatti (*C. jambheri* x *C. karna*) are popular root stocks for Blood Red in Punjab. Rangpur lime is popular rootstock for mandarins and sweet oranges.

Planting: In citrus planting should be done at an spacing of 5-8m. Its depend upon the species and rootstock like trifoliate orange. For acid lime spacing should be maintain 5x5m. the best time for planting is rainy season and February-March. Befor planting the plant 60x60x60 cm pits are prepaired and filled with the soil FYM and carbofuron.

Manure and fertilizers: The bearing tree should be given the fertilizer dose at the ratio of 550gm: 370gm: 550gm NPK/plant/year, depending upon its performance. Graded dose of fertilizers can be applied from 1st year to 10th year. Fertilizers should be applied in a ring 30-40 cm wide just below the canopy of the tree at a distance of at least 1-2 m from the trunk. For improving the fruit quality the application of 0.2% magnasium sulphate and 0.1% boric acid in the month of April.

Irrigation: Irrigation requirements depend upon soil and weather conditions. Water should never come in direct contact with the trunk of the tree, for this reason double ring or check bund method is best suited for this crop. Presently drip irrigation is becoming popular which helps in saving of irrigation water. Stopping irrigation 1 or 2 months prior to flowering is beneficial to the crop, till the tree withers and drops half of its leaves. In summer crop should be irrigated at 8-10 days interval and in winter 12-15 days interval.

Harvesting and yield: Being non-climacteric should be harvested only after full maturity of the fruits. Sweet orange takes 9-12 months for maturity. Harvesting seasons are Dec.-Feb. in North India, Oct.-March in South India, Nov.-Jan. (Ambebahar), March-May (Mrigbahar) in central and western India. Yield varies from 500-2000 fruits/tree depending upon the variety, agro-climatic conditions and age of the tree.

2. Mandarin

Mandarin occupies 50% of citrus area in India. Nagpur santras are grown in saptura hills of central India, hilly slopes of Darjeeling and Coorg. Kinnows are found in Rajasthan, Punjab, Haryana, H.P, J & K and U.P, Wynad, Nilgiri, Palni and Shevoroy hills of South India, hills of Meghalaya, Mizoram, Tripura, Sikkim, Arunachal Pradesh and Bramhaputra Valleys.

Climate and soil: Mandarins grow well in frost free tropical and sub-tropical regions with an elevation of 370-1500m, average temperatures of 10ºC-35ºC and an annual rainfall of 100-1200 cms. Hot winds and severe heat during flowering and fruit set result in heavy loss of yields. Kinnow cultivar requires light chilling for good yields. Mandarins grow in a wide variety of well drained soils but light loamy soils are best. It is grown in soil pH range of 4.5-9.0 depending upon region and type of soil, but 6.0-8.0 is an ideal pH.

Varieties

1. **Coorg**: Fruits medium to large, bright orange, obvate globose, base depressed, rind thin (matures during Feb.-March) found in - Coorg and Wynad.
2. **Khasi**: Fruits depressed, globose to ovate, orange yellow to orange, smooth surface even based or short necked, grow in Assam, Meghalaya, and North Eastern states.
3. **Nagpur**- (Ponkan) Fruits medium, subglobose, coloured, smooth surface, base slightly drawn out (matures Jan.-Feb.). Nagpur is the most popular variety of mandarin, grwon in satpuda hills in Maharashtra.
4. **Kinnow**: (King mandarin X William leaf) fruit medium obvate, base flattened, deep orange yellow, thin rind adhered to segments, irregular bearing, matures in mid January, man made fruits crop and first introduced to India (Punjab) in 1959.

Propagation: By seeds using nuclear seedlings and T budding on Rangpur lime, Rough lemon, Cleopatra, Troyer Citrange, Kharna Khatta rootstocks.

Planting: Monsoon is the best season for planting and pit size can be of 50 cm^3 and spacing is 4-6m. Kinnow on Troyer Citrange can be planted at 1.8m spacing (3000 plants/ha).

Manuring: A fertilizer dose of 450 g of nitrogen, 450 g phosphorous and 900 g of potassium per plant in two split doses 1st dose in May and 2nd after harvest May be given from the beginning. As mandarin also shows micronutrient deficiencies, a composite spray of the following is recommended.

Irrigation: Irrigations may be at 10-15 days interval in winter and 5-7 days in summer. Drip irrigation was found to be more beneficial. Water stress during growth, flowering, fruit set and development should be avoided. In south India, mandarins are grown under rainfed conditions in heavy rainfall areas.

Crop regulation or bahar treatment: Under south and central Indian conditions, mandarin blooming flower three time in a year

February : Ambe bahar, (Fruit drop more serious problem)

June : Mrig bahar (Its preferred for cultivation)

October : Hasta bahar (Mostaly attack of insect pest)

Flower regulation is done by withholding water a month or two before flowering till the plants wither and drop some of their leaves. Then they are manured and irrigated which results in profuse flowering.

Harvesting and yield: Mandarin being non-climacteric should be harvested at right stage of maturity only. Its starts bearing from 4 years but commercially yields can be obtained from 10-12 years. Harvesting season starts from August and extends upto April depending upon the locality and the variety grown. Yield may vary from 500-1000 fruits/plant. Dipping the fruits in 50 ppm ethrel develops golden yellow colour in 5 days. It can be stored for many months at 8-10°C and 85-90 % RH. Under room temperature, they can be stored well for 3-4 weeks.

3. Limes and lemons

India is the largest producer of acid limes in the world, but ranks fifth in the production of limes and lemons. Acid lime is the third important citrus fruit after orange and mandarins. It is cultivated in almost all states Andhra Pradesh, Maharasthra, Tamil Nadu, Karnataka, Gujarat, Bihar and Himanchal Pradesh, are the major ones. Lemons are commercial grown in Punjab, Rajasthan and Uttar Pradesh.

Origin: Limes–India; Lemons–Malaya

Climate and soil: Acid lime grows well in areas which are free from frost, strong winds and are warm and moderately humid. Can grow upto 1000m elevation with annual rainfall not exceeding 750 mm. Sweet limes can tolerate frost better than acid lime. Rangpur lime grows well under areas with 20-30°C. Lemons can be grown in both humid and semi humid areas upto 1200m elevation. Lime grows well in deep (2-2.5m) well drained soils rich in organic matter. Optimum pH is 6.5-7.0 it cannot grow in water logged soils, alkaline soils with high lime content. Lemons grow better in shallow soils also if well drained.

Limes varieties: Kagzi lime is the best most popular one. Some of the improved varieties of limes are: Pramalini, Vikram,(off season) Sai Sarbat: (supposed to be canker free and prolific bearers)

1. **Chakradhar:** Seedless strain of acid lime, PKM-1 heavy yielder.
2. **Selection 49:** Tolerant to canker, tristiza and leaf miner. It is also a prolific bearer.
3. **Seedless lime:** Fruits are oblong, prolific bearer. Tahiti lime- plants nearly thornless, leaves much larger of different shapes, fruits also much larger, triploid.
4. **Mithachikna (Thin rind) and Mithotra (Thick rind):** Two varieties of sweet lime with less acid, sweet fruits and are mostly used as stocks. Rangapur lime is grown for root stock purpose.

Lemons varieties

1. **Eureka:** Fruit elliptic to oblong, base necked and heavy yielder.
2. **Lisbon**: Fruits oblong, apex nippled.
3. **Villafranka**: Fruits oval-oblong, apex pointed, bear more seeds.

4. **Lucknow seedless** (Assam lemon): Fruit oblong, apex nippled, seeds few or absent.
5. **Nepali round:** Nipple has a scar.
6. **Kagzikalan:** Fruit spherical, apex slightly nippled.
7. **Plant lemon 1:** Selection of Kagzi kalan tolerant to canker tristiza and die-back, selfcompatible.

Propagation: Acid lime is propagated by seeds selecting the nucellar seedlings. But budding on Rangpur lime preferable. Sweet limes can be propagated by sub-terminal stem cuttings or layerings or budding. Lemons by budding on Trifoliate orange or Rangpur lime (*Citrus limonica*).

Planting: It can be done before or after rains during the rainy season depending upon amount of rainfall. Spacing should be 4-6m for limes, and 6-8m for lemons. Pits can be of 1 m^3 and filled with mixture of soil and manure.

Manuring: Fully grown lime may be given 5 kg FYM, 900g N, 250g P and 500g K per plant per year. Total P, half of N and total K after rains and remaining dose should be given after flowering.

Irrigation: Limes require more water; irrigation may be given if top 25cm soil becomes dry. Check bund system is best as it will not allow the contact of water with the stem and drip irrigation improves yield.

Harvesting and yield: Harvesting season extends from January-September depending upon the locality for acid lime and May for lemons and June-Aug for Rangpur limes. Lime yields 2000-5000 fruits/plant and lemons yield 1500-3000 fruit/plant. Acid limes can be stored for 6-8 weeks at 8.3 to 10.0^{0}C and 85.9 % RH and Lemons for 8-12 weeks at 7.2-8.6°C and 85-90 % RH.

Physiological disorders

1. **Fruit drop:** This is serious problem of fruit drops in India where the two types of fruit drops occur. The first drop is June drop which is more common in May-June and second drop is pre-harvest drop which is more common in October-March. The application of 2, 4-D 8 ppm in the month of April can control the fruit drop. The spray of NAA 80 ppm in the month of September-first week of October also minimizes the fruit drop. Fruit drops may be causes by climatic factors, improper water management, lack of nutrition etc.

2. **Granulation:** Mosambi, Hamlin and Blood Red cultivars are mostly affected by this disorder. It is drying up, becoming hard with greyish colour and enlargement of the juice vesicles with increase in pectin, lignin, etc. resulting in reduction of the juice content.This occurs more in young vigorous trees

than old trees. The granulation is mainly affected by climate (humid), cultivars, rootstocks, mineral nutrition, enzymes and plant growth regulators and crop load.

It can be control by spraying lime, ZnSo4, Bordeaux mixture or ZnSo4 + CuSo4 (0.5%), 200-500 ppm lead arsenate. Early picking and reduced irrigations is also helpful.

3. **Fruit cracking:** This disorder is more common in lemon but some time it is in malta also. It is mainly due to moisture fluctuation in soil (Irregular interval of irrigation). It can be minimize by application of potasium sulphate @ 4g/ liter of water in the month of April, May and June.

4. **Citrus decline:** It occurs in mandarin fruit crop. It may advance in badly maintained orchards in ill drained soils and also due to malnutrition and insects and pests. Tristeza a very serious viral disease causing various deficiency symptoms in leaves like leaf falling, root decay, dieback of shoots and finally the death of the tree. Trees normally produce heavy crops before dying. Though, appropriate control measures are not available as a means of rejuvenating the declining plants, the following measures may be followed:

- Removing all dead wood before new growth starts and spraying Bordeaux mixture and covering all cut surfaces with Bordeaux paste.
- Applying more organic manures to the plants
- Regularly following the recommended plant protection and spraying 0.3% zinc sulphate and Bordeux mixture whenever new growth appears.
- New plantations should be with plants grafted on Rangpur lime which is reported to be highly tolerant to tristeza.

Insect and pest

1. **Citrus psylla** (*Diaphorina citri*): This is the sucking type pest and they suck the sap from tender part. It can be control by the application of monocrotophos 0.05% or phosphamedan 0.2% or diamethoate 0.05%.

2. **Lemon butterfly** (*Papilio demolius*): This disease is most dangerous in nusery stage. Insect eating the leaves and in seviour condition plants are almost leafless. Insect infestation is more in the month of April-May and August-October. It can be control by application of carboryle 0.2% or a biotic agent like *Bacillus thuringiensis* 0.05% solution.

3. **Leaf minor** *(Phyllocnistis citrella)*: This disease is most dangerous in nursery stage as well as young plant. In seviour condition insect eat the skin of the leaves. It can be control by 0.05% solution of monocrotophos.

4. **Aphid** (*Toxoptera citricidus, Toxoptera aurintii*): In India beside two species of aphid the *Aphis gossypii* and *Myzus persicae* are the equally important. This is the sucking type pest and infestation is more in the month of December-March this insect is suck the sap from leaves and inflorescence. This is also the transmitting agent of tristeza virus. It can be control by 0.05% solution of nicotin sulphate or 1% solution of neem oil.

Disease

1. **Tristeza :** It is a viral disease and transmited by aphid and specific symptoms of tristeza is pitting of inner phase of the bark or trunck seviour pitting develop in lime species Citron while sweet orange, mandarin, rough lemon and sweet lime the stem pitting is usually absent. Affected tree developed various diffieciency symtoms like leaf falling, root decay, dieback and ultimately plant die.

   It can be control by the use of tolerant rootstock like rough lemon, sweet orange, rangpur lime, trifoliate orange, by the use of bud wood free from viruses. It can also be control by checking the aphid population through insecticide.

2. **Greening:** It is caused by mycoplasma. This is the second more serious disease of citrus. The affected plant show chlorosis of leaves similar to zinc difficiency, thickning of leaf blade, saltaning of internode, off season flowering, leaf shading and dieback. The disease spread through infected bud wood. Insect vector i.e. known as Citrus psylla (*Diaphorina citri*) is the main transmiting agent.

   It can be control by the removal of the infected tree and controlling the insect vector by the use of insecticide.

3. **Citrus canker** (*Xanthomonas campestris pv. citri*): It is wide spread in India. It is a serious problem in kagzi lime. The disease affect all plant part like leaf, twig, fruits etc. Canker lesion appears as minute water soaked roundish spot which in large slightly and turn brownish corcky. It can be control by the pruning of the affected part and spraying of 1% solution of Bordeux mixture on emerging flush or application of streptomycin sulphate 500 ppm.

4. **Phytopthora rot :** This is also known as fruit rot, coller rot, crown rot, root rot and phytopthora gummosis. There are three species of phytopthora viz. *Phytopthora parasitica, Phytopthora citropthora, Phytopthora palmivora* has been reported as the casual organism of this disease. Fungus is soil born it's growing well under high moisture condition. It can be control by using the resistance root stock and provide good drainage facility.

28

Jackfruit

Botanical Name	:	*Artocarpus heterophillus*
Family	:	Moraceae
Chromosome No	:	2n = 56
Type of Fruit	:	Sorosis
Edible Portion	:	Bract and Perianth
Origin Place	:	India

Origin and history: Jackfruit is a common and popular fruit crop of the low land tropics of eastern and southern parts of India. This fruit tree grows well without much care. It is the largest edible fruit (20-40kg), in Srilanka a fruit of about 120 kg was harvested and created the world record. The fruit is also called as poor man's food in eastern and southern part. The tree has considerable potential, but it is yet to receive due attention in India it is not generally cultivated in regular plantations and is mainly grown in homestead for domestic uses and as a shade tree in coffee or cardamom plantations. The largest area of jack fruit is in Assam, jack fruit can grow in a wide variety of soil but in grow well in rich, deep alluvial soil.

Species and varieties

The *A. hetrophyllus* previously known as *A. integnifolia*. The other related jack fruit species are :

- *A. altilis* (Bread fruit): Small sized round fruit, found in Tamil Nadu, Assam, Maharashtra and Kerala.
- *A. lakoocha* (Monkey Jack): Small edible fruits, it grows wild in Assam, West Bengal, Uttar Pradesh & Western ghats.
- *A. hirsute* (Aini): It is a semi-wild edible species, native to India.
- *A. champeden*: Edible fruit with strong odour like durian.

Important variety- Gulabi, Champa, Hazara, Rudrakshi, or Singapore Ceylone Jack, Kathal, Kathali, Baramasi, Muttam Varikka, Monkey Jack and Golden Nuqget, Black Gold, Lemon Gold are exotic varieties.

Narendra Dev variety : N.J.T.-1, N.J.T.-2, N.J.T.-3, N.J.T.-4, develop for table purpose and N.J.C.-1, N.J.C.-2, N.J.C.-3, N.J.C.-4, suitable for culinary purpose.

Soil and climate: Proper drainage is ideal for cultivation. The fruit tree is very well suited for dry land conditions. It can be grown under wide range of soils rich deep and well drained soil is good for its cultivation. The fruit tree comes up well under humid and warm climate of hill slopes, arid warmer plains of south India. It is grown up to 1500m from mean sea level and sensitive to frost and drought.

Propagation: Jack fruit can be propagated by both sexual and asexual method. In sexual method seed should be sown immediately just after extraction since they lose their viability during storage (This is due to recalcitrant in nature). The socking of seed in 25 PPM, NAA for 24 hour improved their germination and seedlings grow. In asexual method air layering budding, inarching, grafting, forkert budding, chip budding, patch budding is more common. The best time for budding is in the month of June and for grafting April is the most appropriate time.

Planting: Jack fruit is planted as a shade tree in coffee garden or as mixed plant, as avenue plant. To have a regular orchard like mango the spacing 10x10 m or 12 x 12 m. is followed, pit size, 1x1x1 m. and planting time is the last week of June-August.

Irrigation: Water requirement of a jackfruit tree varies according to climatic conditions. The very first irrigation should be done just after planting and further irrigation should be given as per requirement. In general the first 2-3 year regular irrigation is needed till the root sytem has penetrate deep.

Manure and fertilizers: One year old plant require 10 kg F.Y.M, 100gm N, 50 gm P_2O_5 and 40gm K_2O same amount arising year by year.

Flowering, floral biology pollination and fruit set

Tree start bearing from 7th -8th year after planting. Flowers generly appear in the month of December and continue up to the March, fruit ripen in summer. Tree is monocious inflorescence is spike, which is covered with the 2 spike lets. Sex of the spike can esealy identify when it is small. The surface of the young male spike is smooth while in case of female spike it is glabrous. Jack fruit is anemophyllus in nature and fruit is sorosis, seeds of jack fruit is vivipary in nature. The fruits of jack fruit are cauliflorae in nature.

Insect and pest

Shoot and fruit borer (*Diaphania caesalis*): This insect bore the tender shoots and bud. It can be control by application of phasphamedan @ 0.2%.

Aphid *(Greenidia artocarpi and toxoptera aurantii)*: This is a sucking type pest which suck the sap from leaves, twig, flowers and small fruits the affected plants growth is arrested and adversally affected the yield. It can be control by application of monocrotophas @ 0.05%.

Bug (*Cosmoscarta retata)*: This is a sucking type pest. Insect suck the sap from tender branchage the affected part show yellowing the leaves, arrest the growth and ultimately low yield. It can be control by application of dimethoate @ 0.03% and insects are collected and destroy.

Disease

Pink disease (*Botryobisidium salmonicolor*): In this disease the white and pink spot developed on the branchage and in seviour condition spot in large, bark ruptured and finaly plant growth is arrested. This can be control by the cut portion should be removed, the cut portion should be pasted by bordeux paste and spray of Bordeux mixture 5:5:50.

Fruit rot or soft rot *(Rhizopus artocarpi):* This disease is more common in most humid areas in this problem fruits rotting before maturity and affected fruits are drop resulting low yield. It can be control by application Dithen-M-45 @ 0.2% solution or bavistine @ 0.05% solution at 15th day interval during fruit growth.

Leaf spot (*Phylostricta artocarpine*): In this problem brown spot appear in the margin of leaves and in seviour condition leaves become dry. It can be control by copper oxichloride @ 3g/liter or mancozeb @ 2g/ liter in rainy season.

Harvesting and yield: The tree attain its peack bearing stage in about 15-16 year after planting (maximum fruit) at this day normally a tree bear up to 250 fruit annually with annual fluctuation of yield the weight of fruit. It is also varies depending on the type of variety. On an average about 40-50 tone fruits/ha could abtained.

29

Temperate Fruits

APPLE

Botanical Name	: *Malus domestica*
	: *Malus pumila*
Chromosome No	: 2n = 34
Family	: Rosaceae or Promoidae
Type of Fruit	: Pome
Edible Portion	: Flashy Thalamus
Origin Place	: South West Asia

Origin and history: Apple is a "King of temperate fruits" and it's belonging to family roasaceae. Plants are deciduous in nature. The apple fruits are well known for its sweet, pomaceous fruits. Apples have been grown for thousands of years in Asia and Europe, and were brought to North America by European colonists. Apples have religious and mythological significance in many cultures, including Norse, Greek and European Christian traditions. Apple is the premier table fruit of the world. It is typical temperate fruit tree, about 40 percent of the world's supply being produced in China. Major apple growing countries are U.S.A, Italy, west Garmany, France, Japan and China. In India a large number of apple variety were introduce in 1887 by Mr. Alexander Coutts in his orchard called hillock head in Mashobra near Shimla, Himachal Pradesh. It was also introduce in Bangalore and Nilgiri hills (Tamil Nnadu). But it was destroy by woolly aphid in 1897. In India apple is commercially cultivated in Himachal Pradesh, Uttar Pradesh, Uttarakhand, Jammu and Kashmir and Arunanchal Pradesh.

Soil and climate: It can be grown in all type of soil it perform well in drained, fertile, deep soil with p^H range from 5.8-6.2 and soil should be free from hard pan.

For optimum growth and fruit fullness apple require 100-125 mm rainfall and 21.10-26.70^{0}C temperature for fruit setting. For normal growth of the apple require more chilling. Improper chilling cause more damage to apple flower buds and high winds are not desirable for apple cultivation.

Propagation: Apple can be propagated by sexual and asexual method. Seed propagation is commonally practiced for raising the seedling for rootstock purpose. For raising seedling plants, the seeds are stratified in sand at a temperature of 4-7^{0}C for 60-90 days. Seed after stratification are sown in February-March at a distance of 5 cm in rows 10 cm apart. In asexual method tounge grafting and shield budding is commonly practiced and the best time for propagation is in the month of June.

Pollination and fruit set: In apple there are two types are variety are found i.e. diploid and triploid. The diploid varieties are self fruitfull because they have planty of good pollen. The triploid varieties are self unfruitfull and are having high male sterility and requires pollinizer. Like Red Gold, Golden Delicious apple crop are cross pollinated in nature and require 6-8 behave for proper pollinated. The crab apple (*Malus baccata*) has been found to be very effective polynizer for apple cultivar.

Species: The genus Malus has 25 species and several subspecies of so called Crab apple. The descriptions of some *Malus species* are as follows:

1. *Malus baccata* (L.) Borkh- (Siberian Crab)
2. *Malus sylvestris* Miller- (Wild Crab apple)
3. *Malus floribunda* Siebold- (Flowering Crab)
4. *Malus angustifolia* (Aiton) Michaux- (Narrow leaved Crab apple)
5. *Malus coronaria* (L.) Miller- (American Crab apple)
6. *Malus ioensis* (Alph Wood) Britton-
7. *Malus domestica* Borkh- (Cultivated apple)

Important commercial verities

King of Apple, Red Delicious, Starking Delicious, Rich-A-Red, Red Gold, Golden Delicious, Ambri (This is the Indian origin in J&K), MC-Intosh, Boldwin, Mutsu etc.

Spur types: Red Spur, Red Chief, Starkrimson, Oregon Spur

Low chilling varieties: Tamar, Vered, Tropical Beauty, Maayan, Michal, Schlomit.

Scab resistant varieties: Sir Price, Prima, Liberty, Redfree, Freedom, Macfree.

Delicious varieties or flavouring variety

Lal ambri: Red Delicious x Ambri.

Sunehari: Ambri x Golden Delicious.

Both verities are developing from fruit research station Salimar Kasmir.

Chaubatia Princes: Early Shanberi x Red Delicious.

Chaubatia Anupam: Early Shanberi x Red Delicious.

Both hybrid develop by Horticulture research and training institute, Chaubatia Ranikhet

Ambred: Red Delicious x Ambri.

Ambrich: Rich-A-Red x Ambri.

Ambrayal: Starking Delicious x Ambri.

These hybrids develop from Regional fruit research institute Mashobra, H.P.

Important root stock

In apple one of the premier institute East Malling Research Institute which is situated in England. Develope a series of root stock on the basis of vigour of the plant i.e. M-Series-Malling Series and MM-Series-Malling Marten Series.

1. Ultra dwarf – M-27.
2. Dwarf – M-8, M-9.
3. Semi dwarf – M-26 (M-16xM-9), M-4, M-7, MM-106.
4. Vigourus – MM-104, MM-109, MM-111.
5. Very vigourus - M-12, M-16, M-25, M-793.

Planting: At the time of planting of the tree in orchard proper decision about the selection of varieties, rootstock, tree size and spacing, pollination and planting procedures must be made with understanding. The pits are dug on the well cultivated land during summer in the month of April- May in advance of planting. Generally the dimensions of pit should be kept 1x1x1m. Apple planting is done in square or hexagonal system. The planting distance is kept 5-8m under different conditions.

Irrigation: Water requirement of an apple tree varies according to climatic conditions. The very first irrigation should be done just after planting and further irrigation should be given as per requirement. In general in summer crop should be irrigated at 10-15 days interval and in winter interval should be 15-20 days.

Manure and fertilizers: The doses of manure and fertilizers depends upon soil fertility, type of soil kind, age of the tree, cultural practice and climate. For one year plant 70 gm N, 35gm P_2O_5, 70 gm K_2O and 10 kg FYM is sufficient. The application of compost phasphoras and potas in the month of December-January in fruiting plant. The first dose of nitrogen should be applied 2-3 week before flowering and the second dose of nitrogen should be applied 4 week after flowering. For improving the fruit quality application of 0.1% boric acid, 0.5% zinc sulphate, 0.4% maganees sulphate and 0.5% calcium chloride are beneficial.

Flowering and fruiting : The time of flower bud initiation in apple is closely related to the growth cessation of shoots on which they are set. The period ranging from growth cessation upto the appearance of floral primordial is the period of physiological differentiation of flower buds. During this period differentiation process proceed within the buds, increasing in number of nodes and inducing some physiological change. When the bud reached a certain stage, it may become a flower bud or may terminate its differentiation becoming a leaf bud. The fruit buds are usually developed on spurs, laterally on two or three year old wood. However, buds may also develop on tips of one year old shoots, which are found mostly when dwarfing rootstock are used. The average maximum and minimum temperature around 20^0 and 10^0C, respectively during flowering in April in Himanchal Pradesh resulted in good fruit set. Very low temperature i.e. less than 4.5^0C reduced fruit set by interfering pollen germination and bee activity.

Physiological disorder

Sun scald: It is first appear as light motaling on the greener surface of the fruit darkening become more seviour and ultimately extent to red surface. Scaled usually affect the skin only but in seviour causase it may extend to fruit flesh. The immature fruits are more succeptible to the scald. This due to extream sunlight.

Bitter pit: The symtomps appear as small, dry, brown pockets usually spherical in shape below the skin and also in cortex. The large and immature fruits are more succeptible than othes the heavy dose of nitrogenous fertilizer, exccesive shedding and heavy pruning are reported to favours bitter pit. Mature fruit show black spot in colour skin of the fruit. An exact cause is not known but sprays of calcium nitrate minimize the problem.

Internal cork or drought spot of apple: It is charecterize by brownish streaks radiating into flesh from the core. Internal tissue become brown in colour and is due to boron deficiency and it can be control by application of borax 0.2% and the storage of fruits at 2.2^0C is also helpful in controlling the problem.

Insect pest

Sanjose scale (*Quadraspidiotus perniciosus*): This is the most serious pest of apple. The light infested tree show small grayish scale on the both surface of stem and branches but in heavy infestation the entire surface of the bark is coverd with grey layor of over lapping scales appears as it is spread with wood ashes.

The sanjose scale appears in the apple tree in two main flushes i.e. May and September. Due to its high frequency and rapid rate of multiplication efficient control majors giving higher percentage of mortality are required. During the end of Feburuary and early April it can be control by application of chlorpyrifos 0.02% or phenitrotlıain 0.50%. If neccessory the second application should be applied at the end of early September to early October (post harvest spray).

Woolly aphid (*Eriosuma lanigerum*): This is the suckinfg type pest which suck the sap from stem, twigs and roots resulting in gall formation. The infestation is more during pre mansoon (April-June) and post mansoon (September-November) its depend upon the climatic condition of the areas. It can be control in non bearing plant by application phorate 10-30 gm per tree and for bearing plant application of 0.03% phasphamedan or rogar, in the month of May-June. The application time in non bearing plant is August-September.

Disease

Apple scab (*Venturia inaequalish*): This is the most serious disease in apple cultivation in seviour case plant having defoliation, failure of fruit bud formation, weakning of tree and ultimatally reduce the fruit yield. It can be control by application of captan or mancozeb or carbendazim 0.3%.

Stem black disease (*Coniothecium cometosporum*): The affected portions become jet black in appearance. The fungus enters in the plant through the wound caused by pruning. In seviour conditions the black craks are found in the fruits, bark and branchage become dry, disease infestation is more in July-August. Disease can be control by application of chaubatia paste in the wound portion of the branches and spray of Dithane M-45 at 0.2%.

Fruit rot : Among the various kinds of fruit rots, soft rot (*Penicillium expansum*), bitter rot (*Glomerella cingulata*), pink rot (*Trichothecium roseum*), brown rot (*Monilinia laxa*) causes huge losses to fruit during transportation and storage. It can be control by dipping of fruits in thiobendazol 0.1%, carbendazime 0.1% etc are recommended for the control of post harvest rot.

Harvesting and yield : Apple crop flowering in the month of February-March fruit are harvested when there colour change from green to light yellow the harvesting is done in the month of June-October. After harvesting fruit should be graded on the basis of fruit colour, shape and size. Average yield is about 30-50kg fruit/tree/year.

PEAR

Botanical Name	: *Pyrus communis*
Chromosome No.	: 2n = 34
Family	: Rosaceae (Pomoidae)
Type of Fruit	: Pome
Edible Portion	: Flashy thalamus
Origin Place	: China and Central Asia

Origin and history: Pears are grown in all temperate regions of the world. Cultivars of the European pear, *Pyrus communis*, predominate in Europe, North America, South America, Africa and Australia, and the Sand or Japanese pear, *Pyrus pyrifolia,* is the main cultivated species in southern and central China, Japan and Southeast Asia. Other pears grown widely in Asia include *Pyrus ussuriensis* (Ussuri pear) and hybrids of *Pyrus pyrifolia* and *Pyrus ussuriensis*. Pear is commercially cultivated in china, Japan, Europe, Australia, North and South America. In India it is commercially grown in J&K, Uttarakhand, Himachal Pradesh and Neelgiri hills.

Soil and climate: Pear grown well in deep well drained, fertile and medium textured soil. It is more tolerate to wet soil but less tolerate to draught. pH of the soil should be range from 6-7.5. In pear cultivation water stagnation should be avoided in the field. Sandy soil or soil with high lime contain should be avoided. The highly acidic or alkaline soils are not preferred.

Pear can be grown in wide range of climate. A temperature range from 27-33^{0}C in summer is very congenial for pear cultivation. At the same time a winter temperature of 7^{0}C or bellow for 2-3 month is essential for most of the cultivar. A large number of pear cultivar require about 1200 hours below 7^{0}C during winter to complete their chilling requirement.

Species: In pear 22 species of the genus Pyrus have been described. The important species are either grown for commercial cultivation or for raising rootstock are given below:

1. *Pyrus communis* L.- Common pear or European pear.
2. *Pyrus pyrifolia*- Japanese or Chinese or sand pear.
3. *Pyrus serotina* Rehd- Shiara.
4. *Pyrus pashia* Hamilton- Kainth or Pashia pear.
5. *Pyrus communis* X *Pyrus pyrifolia*- Hybrid pear.
6. *Cydonia oblonga*- Quince A&C

Important verities: On the basis of climate requirement pear have two type of verity i.e.:

1. **European types (*Pyrus communis*):** Where fruit are large in size soft flesh and susceptible to fire blight. The important verities are Bartlett, Beurre-Hardy, Conference, Winter Nellis, Thoumb Pear and Baggugosha.
2. **Asiatic types (*Pyrus pyrifolia*):** This is a low chill variety and there fruit are more crispy (granular) and it is resistant to fire blight. The important verities are Patharrnakh and Twentieth Century.
3. **Hybrid group:** Develop by cross between *Pyrus communis* and *Pyrus pyrifolia*. This is the low chill verity and important verities are LeConte and Kieffer.

Propagation: Pear are mainly propagated by grafting and budding. Shield budding and tongue grafting is more common and the best time for propagation is in the month of June. The rootstocks are raised from the seeds of selected varieties. The seeds of the selected variety should be shown after stratification in the month of February-March, at a distance of 5 cm in row and 10cm in plant. At the time of stratification of the seed, sand temperature is maintained for 4-7^{0}C for 60-90 days.

Planting : At the time of planting selection of the varieties, tree size, spacing, pollination and planting procedures must be made with understanding. The pits are dug on the well cultivated land during summer in the month of May-June in advance of planting. Generally the dimensions of pit should be kept 1x1x1m. The planting distance is kept 5x5m for seedling plant and 3x3m for clonal plant under different conditions.

Irrigation : Water requirement of a pear tree varies according to climatic conditions. The first irrigation should be done just after planting and further irrigation should be given as per requirement. The regular irrigation at 7-10 days interval during April-Augast as per needed. After harvesting tree may be irrigated at 3-4 week interval till they reach leaf shedding stage and there is no need of irrigation during December-January.

Manure and fertilizers : The manure and fertilizer requirement is near to apple requirement. In pear dose of manure and fertilizers depends upon soil fertility, type of soil kind and age of the tree, cultural practice and climate. For one year plant 70gm N, 35gm P_2O_5, 70gm K_2O and 10kg FYM is sufficient. The improvement of fruit quality application of zinc sulphate and ferrus sulphate 0.4-0.5%.

Flowering and fruiting : The flowers are epigenous and the ovary is inferior. The type of inflorescence in pear is corymb. The flowering starts in the last week of February and continues up to third week of March. The fruit is borne on spurs. Fruit setting in pear start by end of March.

Physiological disorder

Sun scald: It is first appear as light motaling on the greener surface of the fruit darkening become more seviour and ultimately extent to red surface. Scaled usually affect the skin only but in seviour cases it may extend to fruit flesh. The immature fruits are more succeptible to the scald. This is due to extream sunlight.

Insect and pest

Sanjose scale: Control by application of chlorphyriphas 0.02% or phenitrothain 0.50%.

Woolly aphid: Control in non bearing plant by application phorate 10-30 gm per tree. For bearing plant application of 0.03% phasphamedan or rogar, in the month of May-June.The application time in non bearing plant is August-September.

Disease

Pear scab (V*enturia pirina*): Infection appears on the young pear shoots. The affected blossoms and petals usually drope, dark brown spots are produced on the fruits. The spot become corky and affected portions are crake. It can be control by application of captan or mancozeb or carbendazim 0.3% and bulk amount of organic manure in the tree.

Fire blight (*Erwinia amylovora*): The affected portions are the branches, flowers and fruits in seviour conditions the affected portion are burn and fruits become black and wrinkled and it can be control by application of streptomycine sulphate 0.02% three spray at weekly interval and the diseased portion should be cut 25cm below the affected part.

Pear decline: It is caused by mycoplasma and transmitted by psylla. In this problem the phloem tubes are checked above the union of root stock and scion shoot that is why the translocation of photosynthate is interrupted and ultimately due to lack of food material plants become dry. It can be control by controlling the transmitting agent by the application of rogar 0.02% and also used the rassistance root stock *Pyrus communis* for the propagation.

Bitter pit: The symtomps appear as small, dry, brown pockets usually spherical in shape below the skin and also in cortex. The large and immature fruits are more succeptible than others the heavy dose of nitrogenous fertilizer, exccesive shedding and heavy pruning are reported to favours bitter pit. Mature fruit show black spot in colour skin of the fruit. Exact cause is not known but sprays of calcium nitrate minimize the problem.

Internal cork or drought spot of apple: It is charecterise by brownish streaks radiating into flesh from the core. Internal tissue become brown in colour and his due to boron deficiency and it can be control by application of borax 0.2% and the storage of fruits at 2.2^{0}C is also helpful in controlling the problem.

Harvesting and yield: Pear is a climacteric fruits which ripen on the tree as well as after harvesting. Fruit size, weight, fruit and flesh colour, TSS and acidity are the major parameter which determines the quality of the fruits. Pear starts fruiting 7 year after planting and on an average 70-80 kg fruit/tree/year and average yield of the fruits is about 25 tones per hectare.

PEACH

Botanical Name	: *Prunus persica*
Chromosome No.	: 2n = 16
Family	: Rosaceae (Prunoideae)
Type of Fruit	: Drupe
Edible Portion	: Epicarp and mesocarp
Origin Place	: China

Origin and history: Peach has a good position amongst stone fruits and is rated as third most important temperate fruit in India. However the cultivated from of the peach came from China. The peach is grown as a commercial and home fruit in most of the temperate countries of the globe. China also grown many peaches and here a large number of forms and types are found. In USA the peach ranks second only to the apple in commercial importance.

Climate and soil: The peach, and its smooth skinned mutant, the nectarine, are primarily grown in temperate zones, between latitudes 30 and 45N and S. The peach flower bud is hardy to about 23^0C to 26^0C which limits its cultivation at higher latitudes. Most peach cultivars require from 100–1000 hours of chilling below 7^0C and they are highly susceptible to early spring frosts. Peach require humid climate with cold winter and dry summer. It is moderate winter hardy and sensitive to low temperature injury. Deep sandy loam soil rich in organic matter is best for its successful cultivation. The appropriate pH of the soil is 5.8-6.8 for its cultivation. Peaches are highly susceptible to water logging and it prefers perfect drainage.

Important varieties

State	Early	Mid	Late
H.P.	Alton, Word-Earliest, Early White, Jayent and Stark red gold	July-Elberta, Kanto-5, Shimizu Hakuto and Sun Heaven	J.H. Hale
U.P.	Early condor, Sun Heaven and Red Heaven	July-Elberta, Alexander Crawford Early	Parrot- Delix, J.H. Hale and Peregrine
J&K	Persevere Quetta, July-Elberta and Saharanpur Prabhat	Alexander, J.H. Hale and Co-smith	

Propagation: Peach is commercially propagated by tongue and cleft grafting and 'T' or ring budding. In hilly area tongue grafting is done in the month of June-February and 'T' budding during the month of May-June.

Planting: Peach should be planted in the end of January when the plants are dormant. Planting should be done after digging 1x1x1m pits refilling them with fertile top soil, mixed 40 kg of well rotten farmyard manure per pit, add 30 g of Lindane 5% dust to each pit to prevent the attack of white ant. Peach plants should be planted at the distance of 6.5 m apart, thus accommodation 225 plants in square system and 259 plants in hexagonal system per hectare. In hilly area peach planted at a distance of 4.5x4.5m. Peach are planting during monsoon period.

Irrigation: Water requirement of a peach tree varies according to climatic conditions. The first irrigation should be done just after planting. In plain area for proper development of fruit size and quality frequence irrigation should be given during April and May. During the month of November when artmospheric temperatures cool down irrigation to peaches should be given at longer duration.

Manure and fertilizers: In peach balance nutritional schedule is important in order to maintained tree health and productivity. Application of 30 gm N, 50 gm P_2O_5, 30 gm K_2O and 10 kg FYM per plant is sufficient. The full amount of FYM, phosphorus and potas it should be applied in the month of December-January and the half dose of nitrogen should be applied in the month of January before the flowering and the rest dose of nitrogen should be applied after one month. Those area where irrigation facility is not proper the full dose of nitrogen should be applied 15 days befor flowering.

Flowering and fruiting: The flowers in peach are perfect, solitary, sessile, in advance of foliage, pink in colour. Flowering starts in the first week of February and continues till end of February and fruit setting starts in the beginning of March.

Insect and pest

Peach leaf curl aphid (*Brachycaudatus helichrysi*): The insects are seviourly infected in the month of March-May in plain area and April-June in hilly area. This is the sucking type pest and sucks the sap of newly born bud and young fruits. In seviour cases infestation may be reapeted after flowering. It can be control by application of dimethoate 0.02%.

Fruit fly (*Bectrocera zonalus*, *Bectrocera dorsalis*): This is the most serious pest in peach the attack of insect at the time of after fruiting this insect is laying the egg inside the fruit after oviposition. Larva is eating the pulp inside the fruit and fruits become roted and fall down. It can be control by application of 0.05% fenitrothion or 0.03% solution of dimethoate.

Diseases

Peach leaf curl (*Taphrina deformans*): Infection of this disease is more in auttum period. In this disease leaf become curl, yellowing and swelling of twig and ultimately plants are die. For controlling the disease application of carbendazime 0.05% at the time of dormancy.

Powdery mildew (*Podosphaera leucotricha*): In this disease white spot appear in the newly emmergase laef and plant growth is adversaly affected. It can be control by application of wettable sulphar 0.3% or carbendazime 0.05% at biweekly interval.

Harvesting and yield: Peach fruits are harvested quickly when ready for harvest. The peach comes into bearing within two years after planting in the field. Fruit are picked when they are still hard as they ripen in storage or transit. The average yield of full grown tree of different varieties varies from 7-10 tone/ha.

PLUM

Botanical Name	: *Prunus salicina*
	: *Prunus domoestica*
Chromosome No.	: 2n = 16
Family	: Rosaceae (Prunoidae)
Type of Fruit	: Drupe
Edible Portion	: Epicarp and mesocarp
Origin Place	: China

Origin, history and distribution: Plums contain a hard pit and thus are classified with other stone fruits in the genus *Prunus* of the Rosaceae. The commercial growing plums are classified into two groups i.e. European (hexaploid) and Japanese (diploid) types. The European plums having chromosome number 48 and Japanese plums are chromosome number 16. European plums are generally better to adapte in cooler regions than Japanese types. Within *Prunus domestica*, several groups of cultivars are recognized such as Green Gage (Reine Claude) types and prunes. The *insititia* subspecies of *Prunus domestica* includes bullaces, damsons, mirabelles and St. Julien types. The greengage, mirabelle and damson plums are used a great deal in the food processing industry. They are processed into jams, jellies, canned fruit, juices (prune juice, for example), and alcoholic drinks such as brandies and cordials. European plums with a sugar content high enough so that they can be dried with the pit intact are referred to as prunes. The term 'Japanese plum' originally was applied to *Prunus salicina* (formerly *Prunus triflora*), but now includes all the fresh-market plums developed by intercrossing various diploid species with the original species. These plums were initially improved in Japan and later, to a much greater extent, in the United States. Most of these plums are consumed as fresh fruit and in many areas of the world these are the predominant plum found in the grocery store.

Plums are commercially grown in Himanchal Pradesh, Jammu & Kashmir, Uttrakhand and Nilgiri Hills. The low chilled varities are commercially grown in Punjab, Hariyana and western Uttar Pradesh.

Soil and climate: Plum can be grown in a wide range of soil but deep, fertile well drained loamy soil with pH range from 5.5-6.5 is good for plum cultivation. Plum require various type of climate. It is grown from subtropical plan to temperate high hill. The chilling requirement of European plum is about 1000-1200 hours and Japanese plum is about 700-1000 hours.

Species: Plum has many species, and taxonomists differ on the count. Depending on the taxonomist, between 19 to 40 species of plum exist. From this diversity only two species, the hexaploid European plum (*Prunus domestica*) and the diploid Japanese plum (*Prunus salicina* and hybrids), are of worldwide commercial significance. The origin of these commercially important species is uncertain but may have involved *Prunus cerasifera* and possibly *Prunus spinosa* as ancestors. Other species of plum variously originated in Europe, Asia and America.

P. cerasifera (Cherry plum)
P. cocomilia (Italian plum)
P. domestica (Species of most "plums" and "prunes")
P. domestica ssp. insititia (Damsons, bullaces)
P. spinosa (Blackthorn or sloe)
P. alleghaniensis (Allegheny plum)
P. americana (American plum)
P. angustifolia (Chickasaw plum)
P. maritima (Beach plum)
P. mexicana (Mexican plum)
P. nigra (Canada plum, Black plum)
P. armeniaca (Apricot)
P. mume (Chinese plum, Japanese apricot)
P. sibirica (Siberian apricot)

Important varieties

Early variety: Jammuny, Titron, Settler, Clothe of Gold, Ramgarh and Meynard.

Mid variety: Howe, Aalu Bukhara, Sharbati, Zardalu Chittidar, Black Champa and Victoria.

Late variety: Late Yellow, Can to 5, Sweet Early, Santa Rosa and Mary Pasa.

Propagation: Plum is generally propagated through hard-wood stem cutting. In case rootstocks are to be used, budding and grafting method are employed for its propagation. Plum is raised on seedling root stock of wild apricot (Zardalu) and Myrobalan B (*Prunus cerasifera*) clonal rootstock. The best time for tongue grafting is in February, for chip budding is in Mid February and 'T' budding is in July-August.

Planting: Plums should be planted in January when the plants are in the dormant condition. The area is laid out and pit of 1x1x1m are dug about a month before planting. The plum plants may be planted 6m apart from row to row and plant to plant in square system, thus accommodating 275 trees per hectare.

Irrigation: Water requirement of a plum tree varies according to climatic conditions. The first irrigation should be done just after planting.The young plum plant upto age of 3-4 year need irrigation at 4-5 days interval during the summer month starting from April-June. Irrigation should be avoided during flowering and fruit setting stage. During April and May frequent irrigation at an interval of 4-5 days should be given for proper development of fruit size and quality. Irrigation should be stope at colour break down stage to avoid heavy fruit drop.

Manure and fertilizers: In plum balance nutritional schedule is important in order to maintained tree health and productivity. One year old plant 70gm N, 35gm P_2O_5, 100gm K_2O and 10kg FYM per plant is sufficient. This dose is increasing praportionaly upto age of 7th year. In case of Zinc difficiency spray of 3kg zinc sulphat and 1.5kg unslaked lime in 500 liter water is beneficial.

Flowering and fruiting: The flowering in Japanese plum varied 11-14 days which started on 5 March and ended on 21 March. The growth pattern of plum followed a double sigmoid curve.

Insect and pest

Plum leaf curl aphid (*Brachycaudatus helichrysi*): The insects are seviourly infected in the month of March-May in plain area and April-June in hilly area. This is the sucking type pest and sucks the sap of newly born bud and young fruits. In seviour cases infestation may be reapeted after flowering. It can be control by application of dimethoate 0.02%.

Fruit fly (*Bectrocera zonalus*, *Bectrocera dorsalis*): This is the most serious pest in peach the attack of insect at the time of after fruiting this insect is laying the egg inside the fruit after oviposition. Larva is eating the pulp inside the fruit and fruits become roted and fall down. It can be control by application of 0.05% fenitrothion or 0.03% solution of dimethoate.

Diseases

Plum leaf curl (*Taphrina deformans*): Infection of this disease is more in auttum period. In this disease leaf become curl, yellowing and swelling of twig and ultimately plants are die. For controlling the disease application of carbendazime 0.05% at the time of dormancy.

Powdery mildew (*Podosphaera leucotricha*): In this disease white spot appear in the newly emmergase leaf and plant growth is adversaly affected. It can be control by application of wettable sulphar 0.3% or carbendazime 0.05% at biweekly interval.

Harvesting and yield: Plums are climatric in nature and their fruits are harvested quickly when ready for harvest. The plum comes into bearing within two years after planting in the field. Fruit are picked when they are still hard as they ripen in storage or transit. Average yield of plum is 40-60kg fruit/tree.